The Planetary Thermostat

Feedback Loops That Stabilize Earth

Team Trex

Published by NobleTrex Press

For permissions and other inquiries, write to:

P.O. Box 3132, Framingham, MA 01701,
USA

Contents

Introduction

Four billion years ago, the sun was a weakling. It burned with a pale, orange lethargy, shining with roughly thirty percent less luminosity than the star that woke you up this morning. By all rights, the infant Earth should have been a frozen wasteland, a sphere of white ice locked in a permanent, lifeless winter. The laws of physics are unsentimental about this: based on the energy output of the young sun, our planet's average surface temperature should have been well below freezing.

But the rocks tell a different story.

Deep in the geologic record, inside the ancient cratons of Greenland and South Africa, we find water-deposited sediments dating back to that dim, early epoch. We find ripples of ancient tides captured in stone. We find evidence of liquid oceans teeming with microbial life. This contradiction is known as the Faint Young Sun Paradox, and it is one of the most profound mysteries in planetary science. How did a planet with a feeble furnace manage to keep the pipes from freezing?

The answer lies in the atmosphere, but not in the way we usually think about it. It wasn't just that the air was different; it was that the air was-and still is-alive. Not biologically alive, perhaps, but chemically and physically dynamic. The Earth was not merely a passive rock bathing in sunlight; it was, and remains, a self-regulating system. As the sun slowly brightened over eons, the Earth stripped greenhouse gases out of its own atmosphere, turning down the insulation just as the furnace cranked up. It maintained a temperature capable of sustaining liquid water for nearly four billion years, despite astrophysical chaos, volcanic apocalypses, and the shifting of continents.

We live inside a miracle of engineering. But unlike the thermostat on your wall, which relies on a simple sensor and a microchip, the planetary thermostat is built from crumbling mountains, deep ocean currents, melting ice sheets, and the respiration of trillions of trees. It is a Rube Goldberg machine of staggering complexity, composed of interlocking loops that push and pull against one another to maintain the delicate equilibrium we call climate.

These are the feedback loops. They are the invisible gears that stabilize Earth. And right now, for the first time in human history, we are jamming a wrench into them.

To understand the moment we are in-a moment of feverish heat records and shifting baselines-we have to move beyond the simple idea of "greenhouse gases." Yes, carbon dioxide is the trigger. But CO_2 is merely the hand that pushes the first domino. The actual climate we experience-the storms, the droughts, the pleasant spring days-is determined by how the system responds to that push.

This is where the story gets fascinating, and admittedly, a little frightening. The Earth is an amplifier. If you warm the planet slightly, it doesn't just get slightly warmer. The warming triggers a cascade. The air, now warmer, sucks more moisture from the oceans. Water vapor is itself a potent greenhouse gas, so this extra humidity traps even more heat, which causes even more evaporation. This is a positive feedback loop-a vicious cycle.

Simultaneously, the white shield of the Arctic melts, replacing a mirrored surface that reflects sunlight with dark water that absorbs it. The planet gets darker, absorbs more energy, melts more ice, and warms further.

These mechanisms are the reason climate change is not a linear problem. If the Earth were a simple rock, doubling the carbon dioxide might warm us by a degree or so-manageable, perhaps. But because of these engines of amplification, that same push threatens to warm us by three, four, or five degrees. We are dealing with a system that has a gain factor. It listens to our input and shouts it back through a megaphone.

However, the story is not entirely one of runaway destruction. If it were, life would have been extinguished long ago. There are opposing forces at play-virtuous cycles that dampen change and seek balance. A warmer planet radiates more energy into the vacuum of space, a frantic attempt to cool itself off. The oceans act as a colossal battery, absorbing ninety percent of the excess heat and burying it in the abyss, buying us precious time. The biosphere inhales carbon, turning our emissions into wood and soil.

This book is an investigation into that tension. It is a tour of the machinery that keeps Earth within the "Goldilocks zone"-not too hot, not too cold.

We often talk about the climate as if it were a static backdrop to human history, a theater stage that stays the same while empires rise and fall. In reality, the stage is a raft on a storm-tossed sea. We have been lulled into a false sense of security by the Holocene, the relatively stable ten-thousand-year summer that allowed humanity to invent agriculture, build cities, and write books like this one. But that stability was never a guarantee; it was a dynamic equilibrium, a tightrope walk performed by a planet balancing a dozen different spinning plates.

Why does this matter to you, right now, as you hold this book?

Because the thermostat is lagging.

One of the most critical concepts we will grapple with is *inertia*. When you turn the shower handle, the water doesn't get hot instantly. The water has to travel through the pipes; the heater has to do its work. The Earth operates on a similar, though much vastly slower, delay. The warming we see today is the response to carbon emitted decades ago. The carbon we emit today locks in the warming our children will experience. We are currently baking a future that hasn't yet come out of the oven.

Understanding these feedbacks changes how you see the world. When you look at a cloud, you won't just see a fluffy decoration; you will see a battleground between reflection and insulation, a chaotic variable that keeps supercomputer modelers awake at night. When you see a forest fire, you won't just see a tragedy for local wildlife; you will see a carbon sink flipping into a carbon source, a guardian becoming an arsonist. When you look at the ocean, you will see a conveyor belt moving heat from the tropics to the poles, a circulation system that dictates the weather patterns of entire continents.

We start our investigation by looking at the ledger. Before we can understand the complex loops, we must understand the basic accounting of sunlight. We will examine the sheer magnitude of the energy pouring down on us and the delicate physics of how Earth sheds that heat back into the dark. It is a game of balance where a difference of less than one percent determines the fate of civilizations.

From there, we move to the amplifiers. We will explore the physics of humidity and the "lapse rate"-the vertical temperature profile of the atmosphere that acts as a release valve for heat. We will see how the very structure of the air dictates how much warming we endure.

Then, we turn our gaze to the surface. We will look at the cryosphere-the world of ice-and the albedo effect. This is one of the most intuitive yet dangerous feedbacks. We are effectively peeling the reflective foil off our planet, exposing the dark, absorbent material underneath. We will look at the geological history of the "Snowball Earth" events, times when this feedback ran amok and froze the planet to the equator, proving that the thermostat can break.

But the Earth is wet, and it is deep. We will dive into the role of the global ocean, the great moderator. We will trace the thermohaline circulation-the "global conveyor belt"-and ask what happens when the pumps slow down. We will see how the ocean creates a deceptive calm, hiding the true extent of our energy imbalance in its deep, dark waters.

Life itself is a player in this game. We will explore the biosphere's "green lung." Plants are not passive decorations; they are active geo-engineers. From the microscopic plankton that seed clouds with their chemical exhaust to the vast boreal forests that sweat water vapor into the sky, biology is intimately tied to physics. We will discuss the frightening possibility of the "permafrost time bomb," where ancient frozen microbes wake up and begin belching methane, potentially wrestling control of the climate away from human influence entirely.

Finally, we arrive at the most human of questions: How hard can we push? We will examine the concept of tipping points-thresholds of no return. Systems like the Amazon rainforest or the West Antarctic Ice Sheet have breaking points. Cross them, and the feedback loops take over, driving the transition to a new, unrecognizable state regardless of what we do next. This is the realm of hysteresis, where the path backward is far more difficult than the path forward.

There is a grandeur to this machinery. When scientists like James Lovelock first began discussing the self-regulating nature of the biosphere in the 1970s, some dismissed it as mysticism. But as our understanding of Earth Systems Science has matured, we have replaced the mysticism with hard mathematics. We can now quantify the coupling strength between the soil and the air. We can measure the albedo of melting ice with satellites. We have mapped the gears.

Yet, there is a profound vulnerability in this knowledge. For billions of years, the feedback loops operated on geological timescales. The "silicate weathering thermostat"-the process by which rain washes carbon out of the air and locks it in rock-takes millions of years to adjust the temperature. It saved the Earth from the brightening sun, but it is far too slow to save us from ourselves. We are dumping carbon into the atmosphere at a rate that outpaces the natural removal processes by orders of magnitude. We are demanding a sprint from a system built for a marathon.

This is not a book about doom. It is a book about mechanics. To fix a machine, or at least to stop breaking it, one must first respect how it works. We are currently operating the spacecraft Earth without reading the manual. We are pulling levers labeled "Aerosols" and "Carbon Cycle" without fully understanding how they connect to the dials labeled "Temperature" and "Sea Level."

The planetary thermostat is resilient, yes. It has recovered from asteroid impacts and super-volcanoes. But "recovery" in geological terms takes ten thousand generations. We don't have that kind of time.

As we peel back the layers of the atmosphere and plumb the depths of the oceans in the chapters to come, keep one thing in mind: We are not external to this system. We are not observers looking through glass. We are inside the feedback loop. Our choices, our economies, and our technologies are now as much a force of nature as the orbital tilt or the solar cycle.

The thermostat is shaking. The needles are redlining. It is time to open the panel and see what makes it tick.

Chapter 1

The Goldilocks Equation

Earth hangs suspended in a cosmic void, caught between the searing intensity of the Sun and the freezing vacuum of space. This chapter dismantles the planetary machinery that maintains our liquid water habitability, revealing how a precise accounting balance between incoming photons and outgoing infrared waves prevents us from freezing or boiling.

1.1 The Solar Stipend

Every climate story starts with a paycheck.

Earth is not warmed from within, at least not in any way that matters for the everyday weather above our heads. Yes, the planet leaks heat from its interior—a faint glow of geothermal energy from radioactive decay and leftover formation heat—but compared with what arrives from space it is pocket change. The serious money comes from the Sun, delivered as a steady stream of electromagnetic energy that has crossed 150 million kilometers of vacuum to reach the top of our atmosphere.

That incoming stream is remarkably constant by human standards. The Sun has moods (sunspots, flares, the slow pulse of the 11-year solar cycle), but the changes are small relative to the baseline. For the basic accounting of why Earth is warm at all—and why it can become warmer or colder—it is enough to treat the Sun as a stable benefactor and ask a deceptively simple question: *how much energy do we receive, on average, per square meter of the planet?*

The solar constant and a historical detour. The quantity that climate scientists start with has a crisp, almost austere name: the *solar constant*. It is not truly constant in time, but it is constant enough that the name stuck. It means the power from sunlight falling on a surface held perpendicular to the Sun's rays at Earth's distance from the Sun, measured at the *top of the atmosphere*. Power per unit area has units of watts per square meter, $\mathrm{W\,m^{-2}}$, and the solar constant is about

$$S_0 \approx 1361\ \mathrm{W\,m^{-2}}.$$

That number is easy to take for granted today because satellites can measure it above the blurring and filtering effects of the air. But the urge to quantify it is older than the space age and has a distinctly human flavor: if sunlight is the planet's income, people wanted to know the size of the deposit.

In the early nineteenth century, the French physicist Claude Pouillet tried to estimate this solar income using an instrument of his own design, a *pyrheliometer*, essentially a carefully constructed absorber whose temperature rise could be linked to the incoming solar power. The challenge was not just craftsmanship; it was honesty about all the ways Earth interferes with the measurement. Even at a high mountain observatory, the atmosphere scatters, absorbs, and re-emits radiation, so what reaches an instrument at the surface is already the Sun's gift after a long chain of deductions. Pouillet's estimates were not perfect—no one could fully correct for atmospheric absorption with the knowledge and tools of his time—but the project mattered because it treated sunlight as a measurable physical input rather than a poetic background condition. Later in the century, Samuel Langley improved the situation with his exquisitely sensitive *bolometer* and, just as importantly, with the idea that one should measure sunlight at different air masses (different Sun angles through the atmosphere) to infer what must have been present above the air entirely.

The exact value has been refined with modern radiometers, but the conceptual move remains the same: quantify the stipend before it gets spent, diverted, or taxed.

The disk and the sphere: why we divide by four. A common misunderstanding arises the moment someone hears $1361\ \mathrm{W\,m^{-2}}$. "If that much power arrives," they ask, "why isn't every square meter of Earth receiving it?" The answer is geometry, and geometry is destiny in planetary climate.

Sunlight arrives in nearly parallel rays. Earth intercepts those rays with its *cross-sectional area*, the area of a disk with radius R:

$$A_{\text{disk}} = \pi R^2.$$

That disk is the planet's "shadow" in space—the target that catches the incoming beam. But the energy is then spread, by rotation and atmospheric and oceanic transport, over the planet's entire surface area:

$$A_{\text{sphere}} = 4\pi R^2.$$

Even before we worry about clouds, ice, oceans, deserts, or greenhouse gases, the average incoming solar power per square meter over the whole planet is therefore reduced by a factor of four:

$$\langle S \rangle = \frac{S_0\, \pi R^2}{4\pi R^2} = \frac{S_0}{4}.$$

Numerically,

$$\frac{1361}{4} \approx 340\ \mathrm{W\,m^{-2}}.$$

This 340 $\mathrm{W\,m^{-2}}$ is a global mean at the top of the atmosphere, averaged over day and night and over all latitudes. It is not what you feel on your skin at noon on a clear day. It is the accountant's number: a planet-wide average that makes the later balancing of books possible.

A small, almost comic thought experiment helps. Imagine freezing Earth in space, locking it so that one hemisphere faces the Sun at all times, like the Moon faces Earth. The dayside would roast and the nightside would freeze, but the planet-wide average incoming energy would still be $S_0/4$, because the same disk would still intercept the beam and the same sphere would still be the area over which we could average. The divide-by-four is not about rotation smoothing out day and night (though rotation matters enormously for habitability). It is about the fact that a sphere presents a disk to a distant light source.

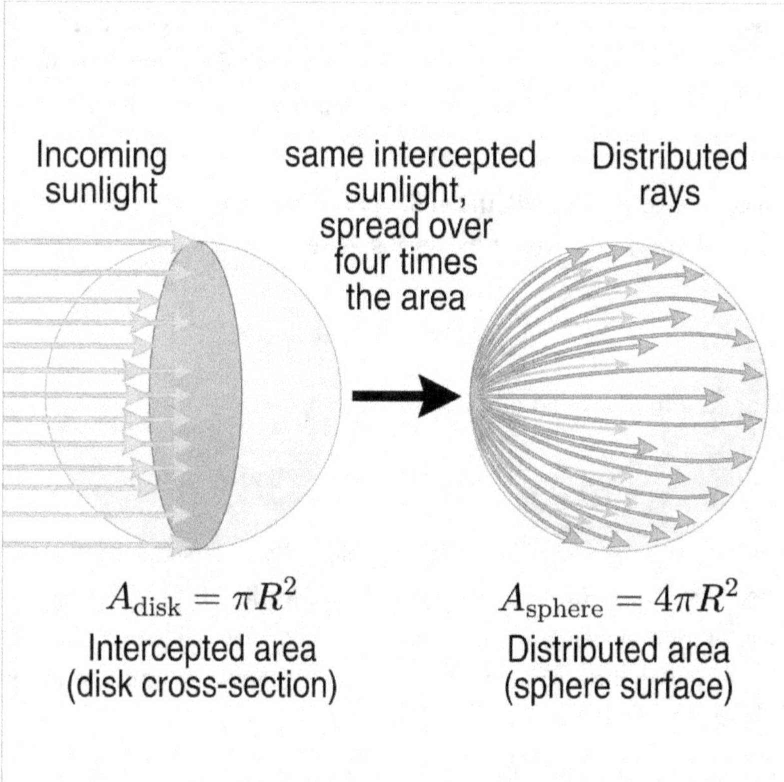

Albedo: the planet's automatic refund. A second deduction comes before sunlight can warm anything: Earth reflects some of it straight back to space. The fraction reflected is called the *albedo*, from the Latin for "whiteness." In the simplest budget, albedo is a single number, α, representing the planet-wide average reflectivity at the top of the atmosphere. Earth's average albedo is about $\alpha \approx 0.30$. That means roughly 30% of incoming sunlight is returned without being converted into heat within the Earth system.

What does the reflecting? Clouds, first and foremost. A bright, thick cloud deck can behave like a mirror for sunlight, sending a large fraction back to space. Ice and snow also matter because they are highly reflective, especially in the visible part of the spectrum where the Sun emits much of its energy. Deserts, salt flats, and even wind-ruffled ocean surfaces contribute, but clouds and ice do the heavy lifting.

Once albedo is included, the globally averaged *absorbed* solar power per unit area at the top of the atmosphere becomes

$$\langle S_{\text{abs}} \rangle = (1 - \alpha)\,\frac{S_0}{4}.$$

Plugging in the round numbers,

$$\langle S_{\text{abs}} \rangle \approx (1 - 0.30) \times 340 \approx 239\ \text{W m}^{-2}.$$

This 239 W m^{-2} is the stipend that actually enters Earth's climate system as usable income. Everything else in climate—winds, storms, evaporation, ocean currents, ice ages, heat waves—is, at root, a story about how that energy is moved around and how it eventually leaves again.

The elegance of this number hides a great deal of texture. Albedo is not fixed; it changes with cloudiness, with the seasonal growth and retreat of snow and sea ice, with dust and aerosols in the air, with the extent of forests, and even with human choices about land use. A planet can become brighter or darker, and when it does, the stipend changes without the Sun doing anything different.

A concrete sense of scale helps here. The difference between absorbing 70% of the incoming solar energy and absorbing 71% may look trivial, but it is a 1% change applied to about 340 W m^{-2} of incoming average sunlight at the top of the atmosphere, which is roughly 3.4 W m^{-2}—a number that climate scientists treat as a substantial perturbation. Albedo is not a decorative parameter; it is a lever.

The stipend is unevenly delivered. The global average is the accountant's tool, but Earth does not live at its average. Anyone who has flown from a winter city to a tropical beach knows that the Sun's gift is distributed unevenly. Two facts drive that unevenness: the curvature of the planet and the tilt of its rotation axis.

At the equator, sunlight tends to strike more directly, concentrating energy onto smaller surface areas. Toward the poles, the same beam is spread out over larger surface areas because it arrives at a shallow angle; the Sun seems to skim the horizon, and the energy is diluted. On top of that, polar regions often sit under higher albedo conditions (snow and ice) precisely where the incoming energy is already weaker.

Then comes the axial tilt, the reason for seasons. Earth's axis is tilted by about 23.5° relative to the plane of its orbit. That tilt causes one hemisphere to lean toward the Sun for part of the year, receiving longer days and higher Sun angles, and away from the Sun for the other part, receiving shorter days and lower Sun angles. The stipend is not merely uneven across latitude; it pulses in time.

This unevenness is not a minor complication; it is the engine of atmospheric and oceanic circulation. If the equator received the same solar income as the poles, there would be far less incentive for the planet to move heat around. Instead, the tropics run a chronic surplus and the high latitudes run a chronic deficit. Winds and currents are the planet's bookkeeping system, transferring energy from where it is overpaid to where it is underpaid.

A human-scale comparison: sunlight as household power. Numbers like 239 $\mathrm{W\,m^{-2}}$ can feel abstract because they are spread over the entire globe and averaged over day and night. To make it more tangible, compare it to a familiar benchmark: a bright incandescent light bulb was once about 60 watts. If you imagine a square meter of Earth, the average absorbed solar income, 239 watts, is like running four such bulbs continuously over that area. But because the average includes nighttime, it implies that the daytime receipt must be larger in most places, which is why a clear noon Sun can feel intense enough to warm your skin quickly.

Or consider human energy use. Global primary energy consumption, averaged over the planet's surface area, corresponds to a fraction of a watt per square meter. Earth absorbs *hundreds* of watts per square meter from the Sun. Human civilization is energetic and transformative, but it operates on top of a far larger solar river. The climate problem is not that humans rival the Sun's stipend; it is that humans can subtly alter the terms under which the stipend is retained and returned, by changing atmospheric composition and reflectivity.

Why the "top of the atmosphere" matters. There is a reason the stipend is defined at the top of the atmosphere. The atmosphere is not a passive window. It scatters sunlight (turning the sky blue), absorbs some wavelengths, and hosts clouds that can dramatically raise the planet's reflectivity. Measuring at the surface mixes the solar stipend with the planet's internal bookkeeping: how much was absorbed on the way down, how much was reflected back upward, how much arrived as direct beam versus diffuse skylight.

In the basic planetary budget, the top of the atmosphere is the clean boundary for the ledger. Sunlight crosses it coming in; reflected sunlight and infrared radiation cross it going out. The complicated middle—the realm of weather, water vapor, aerosols, convection, and clouds—is real life, but the boundary is where the accounting closes.

A useful mental picture is a building with a glass atrium. You can measure how much money the building receives by looking at deliveries at the door, not by walking through the offices and trying to infer what arrived from what you find scattered on desks. The top of the atmosphere is the door.

Small changes, big consequences. Once the stipend is written down, it is natural to ask how stable it is. Three broad factors can change the incoming solar term in the energy budget:

First, the Sun itself changes slightly. Total solar irradiance varies over the solar cycle by roughly a tenth of a percent. That is small but not negligible for precise studies, and it can leave fingerprints in the upper atmosphere and in some patterns of variability.

Second, Earth's orbit and tilt change on long timescales. Over tens to hundreds of thousands of years, slow gravitational nudges from other planets alter the shape of Earth's orbit, the tilt of its axis, and the timing of the seasons relative to where Earth is in its orbit. These Milankovitch cycles redistribute sunlight in latitude and season, providing the pacing for ice ages. The key word is redistribute: the annual global mean stipend changes little, but where and when the energy arrives changes enough to reorganize ice sheets and oceans.

Third, the planet's reflectivity changes. This is the most climate-relevant lever on human timescales because albedo can be altered quickly by clouds, aerosols, and surface changes. Volcanic eruptions provide a dramatic natural example. When Mount Pinatubo erupted in 1991, it injected sulfur dioxide high into the stratosphere, where it formed sulfate aerosols that reflected additional sunlight back to space. The result was a measurable, temporary cooling of the planet. The Sun's stipend did not shrink; the refund grew.

There is a quiet lesson in this. Earth's climate is sensitive not just to the size of the stipend but to the fine print of how much is accepted and how much is returned immediately. A small change in albedo—a slightly brighter planet—can offset a surprisingly large amount of warming influence elsewhere, at least temporarily. But because clouds and aerosols are patchy, transient, and entangled with atmospheric circulation, they are among the most challenging parts of the climate ledger to pin down with high confidence.

A compact equation with a lot of meaning. For all the complexity of real sunlight on a real, cloudy, rotating planet, the raw income side of Earth's energy budget can be summarized in a single line that is simple enough to fit on a napkin:

$$\boxed{\langle S_{\text{abs}} \rangle = (1 - \alpha)\,\frac{S_0}{4}}$$

with S_0 set by the Sun and Earth's orbit, the factor of 4 set by geometry, and α set by how bright the planet chooses to be.

The elegance of this expression is not that it captures every detail—it does not—but that it places hard boundaries on the rest of the discussion. Whatever the atmosphere and oceans do, whatever storms and clouds conspire to create, whatever feedbacks amplify or damp changes, they are negotiating over a stipend of order a few hundred watts per square meter. When the planet warms, it is not because the stipend suddenly became infinite; it is because the terms of exchange between income and expenditure shifted, leaving a small but persistent surplus in the account.

A planet can survive a lot of weather on a balanced budget. It is the slow drift, the steady miscounting of a few watts per square meter sustained over decades, that accumulates into melted ice, rising seas, and shifting climate zones. The numbers in the stipend are big enough to run the entire machinery of winds and oceans, and precise enough that even small percentage changes can steer the long-term fate of coastlines and crops.

1.2 The Infrared Exhaust

A planet that only collected energy would not stay pleasantly warm for long. It would keep warming until oceans boiled and rocks glowed. Earth avoids that fate because it has a built-in exhaust pipe: it radiates energy back to space.

The key twist is that this exhaust is not visible light. Earth is far cooler than the Sun, and cooler objects glow at longer wavelengths. The Sun's surface, at roughly 5800 K, emits strongly in the visible and near-infrared; Earth, with surface temperatures near 288 K, emits mostly in the thermal infrared. That is why the planet can be bathed in sunlight and yet lose energy in a kind of invisible ember-glow.

A quick everyday example makes the idea visceral. On a calm, cloudless night, the ground often cools faster than the air above it. Gardeners learn this the hard way: a "radiation frost" can form even when the thermometer a meter or two above the soil reads safely above freezing. The soil and leaves are radiating infrared energy to the cold sky, and without clouds to send some of that radiation back down, they can become cold enough for ice to form locally. The exhaust is not an abstract concept; it is something you can wake up to on a spring morning and find on your windshield.

The power of the infrared exhaust is that it is *self-adjusting*. Warm things radiate more than cool things. If Earth warms for any reason, its infrared emission increases, and that increased loss of energy tends to push temperatures back down. If Earth cools, emission decreases, which slows the cooling. This stabilizing tendency is the simplest feedback loop in climate: the *Planck response*, named after the physics of thermal radiation.

Stefan, Boltzmann, and the mathematics of glow. The relationship between temperature and emitted radiation was not obvious in the nineteenth century, when heat and light were still being disentangled as physical phenomena. Josef Stefan, working in Vienna, inferred an empirical law in 1879 by looking at measurements of hot objects and the physics of cooling. Ludwig Boltzmann provided the theoretical foundation in 1884 using thermodynamics and electromagnetic theory. Their result is now one of the workhorse equations of climate science, astrophysics, and engineering:

$$F = \sigma T^4.$$

Here F is the emitted power per unit area (the radiative flux), T is absolute temperature in kelvins, and σ is the Stefan–Boltzmann constant,

$$\sigma \approx 5.67 \times 10^{-8}\,\mathrm{W\,m^{-2}\,K^{-4}}.$$

The fourth power matters. It means emission rises very rapidly with temperature. Double the temperature and emission increases by a factor of $2^4 = 16$. Earth is nowhere near doubling its absolute temperature, of course, but the curvature of T^4 is still the reason the infrared exhaust behaves like a governor: it becomes harder and harder to keep adding heat without the planet radiating more vigorously in response.

Real materials are not perfect emitters. A perfect emitter is a *blackbody*: it absorbs all incident radiation and, at a given temperature, emits the maximum possible thermal radiation with a characteristic spectrum. Many surfaces behave close enough to a blackbody in the thermal infrared that the approximation is surprisingly useful. Earth's surface, in particular, has a thermal infrared emissivity close to one, meaning it emits almost like a blackbody at those wavelengths. The atmosphere complicates things, but the Stefan–Boltzmann law remains the backbone of the accounting.

From solar income to radiating temperature. The stipend arriving from the Sun can be turned into a temperature by demanding balance: in long-term equilibrium, average absorbed solar power at the top of the atmosphere must equal average outgoing infrared power to space. Using the globally averaged absorbed sunlight,

$$\langle S_{\text{abs}} \rangle = (1 - \alpha)\frac{S_0}{4},$$

and equating it to a blackbody emission,

$$(1 - \alpha)\frac{S_0}{4} = \sigma T_{\text{e}}^4,$$

where T_e is the *effective radiating temperature*—the temperature a perfect blackbody would need to have in order to emit the observed outgoing infrared flux.

With $S_0 \approx 1361\ \text{W}\,\text{m}^{-2}$ and $\alpha \approx 0.30$, the absorbed solar is about $239\ \text{W}\,\text{m}^{-2}$. Solve for T_e:

$$T_\text{e} = \left(\frac{239}{\sigma}\right)^{1/4} \approx 255\ \text{K}.$$

In everyday units, 255 K corresponds to about -18°C.

That number is famous because it is wrong in a revealing way. Earth's average surface temperature is closer to 288 K (about 15°C), roughly 33°C warmer than -18°C. The planet radiates to space as though it were a cold object, yet the surface where people live is much warmer. This discrepancy is not a minor detail; it is the signature of atmospheric physics. The infrared exhaust exists, but it is routed through valves and ducts rather than vented straight from the ground to space.

Before worrying about those valves, it is worth seeing how the exhaust alone builds stability into the system. The Stefan–Boltzmann law implies that a small change in temperature produces a proportional change in emitted flux:

$$\frac{dF}{dT} = 4\sigma T^3.$$

At $T_e \approx 255$ K, this slope is about $3.8\ \mathrm{W\,m^{-2}\,K^{-1}}$ (a useful rule-of-thumb number). In plain language: if the effective temperature rises by 1 K, the planet emits roughly $4\ \mathrm{W\,m^{-2}}$ more to space, pushing back against whatever caused the warming. That is the basic radiative brake.

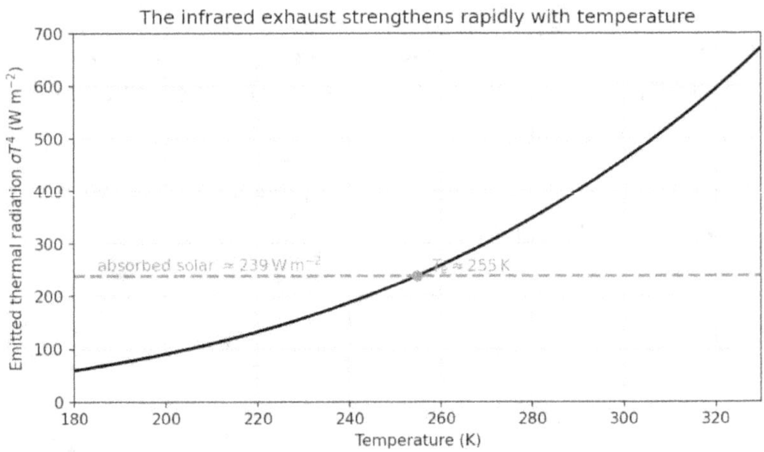

Why $-18°$C *is not the surface.* The effective radiating temperature is a planet-wide accounting convenience, not a thermometer reading from the ground. It refers to the temperature implied by the outgoing infrared flux at the top of the atmosphere. If Earth were a bare rock with no atmosphere and the same reflectivity, the surface itself would need to sit near 255 K to radiate $239\ \mathrm{W\,m^{-2}}$ to space. The actual Earth has an atmosphere that absorbs and emits infrared radiation. That means the radiation escaping to space often comes from higher altitudes, where the air is colder than the surface. The planet can then emit to space as though it were 255 K while maintaining a warmer surface below.

One way to make this feel less like a trick of definitions is to imagine taking an infrared photograph of Earth from space. You would not see the surface everywhere. Over clear skies you might glimpse warmer emission originating closer to the ground; over thick clouds you would see colder emission from cloud tops. In the aggregate, the outgoing infrared looks like it is coming from a patchwork of temperatures and altitudes, and the global mean of that patchwork corresponds to about 255 K. The surface can be warmer because it is not always the part of the system directly venting to space.

The exhaust is not a single pipe. When people first hear "Earth radiates 239 $\mathrm{W\,m^{-2}}$ to space," it is tempting to picture a uniform glow rising from the surface like heat from a stovetop. The real plumbing is more intricate.

Some infrared radiation does escape directly from the surface through a relatively transparent part of the spectrum often called the *infrared window*, roughly between about 8 and 14 μm in wavelength. On a dry, clear night, this window is wide open, and surface cooling by radiation can be strong—which is exactly why deserts can be hot in the day and surprisingly cold at night. Increase humidity, and water vapor partially closes that window. Add clouds, and it closes further, because cloud droplets and ice particles are efficient absorbers and emitters of infrared.

Much of the surface's infrared emission does not go straight to space. It is absorbed by greenhouse gases and clouds, heating the atmosphere locally. The atmosphere then emits infrared radiation both upward and downward. The upward component contributes to the top-of-atmosphere exhaust. The downward component is part of what keeps the surface warm. In energy-budget diagrams of the Earth system, this atmospheric back-radiation is a major term: large in magnitude, but not a violation of energy conservation, because it is paired with equally large upward flows and with non-radiative transfers.

Non-radiative transfers matter too. A great deal of the energy that warms the atmosphere begins at the surface not as infrared radiation but as *latent heat* (carried by water vapor evaporated from oceans and plants) and *sensible heat* (carried by warm air rising and mixing). When water vapor later condenses into cloud droplets, it releases latent heat aloft, helping power storms and convection. This is still part of the planet's bookkeeping: energy acquired from the Sun becomes heat at the surface, is transported upward by phase changes and air motion, and is ultimately radiated away as infrared from higher, colder layers.

In other words, the exhaust comes from the *surface–atmosphere system*, not just the surface.

A stabilizer with a temperament. The Stefan–Boltzmann law makes the exhaust strongly temperature-dependent, which sounds like a guarantee of stability. In a broad sense it is: there is always a pushback. But the strength of the pushback depends on the details of how easily infrared radiation can escape.

Imagine partially covering the exhaust of a fireplace with a damper. The fire can still burn, and hot gases can still leave, but for a given fire intensity the room becomes warmer before the system can shed the same amount of energy through the restricted opening. In climate terms, changing greenhouse gas concentrations or cloud properties can alter the relationship between surface temperature and outgoing infrared radiation. The underlying T^4 dependence is still there, but the temperature that matters for emission to space may shift to higher or colder layers, changing how much extra surface warming is required to restore balance.

This is why the infrared exhaust is both reassuring and unsettling. Reassuring, because the physics supplies a firm negative feedback: warm more, radiate more. Unsettling, because the atmosphere determines *where* the radiating happens and therefore how quickly the exhaust increases as the surface warms. The thermostat is real, but it has settings.

Cooling you can feel: the clear-sky night. Return to that gardener's frost. Clouds are a good laboratory because they change the infrared exhaust in a way that is immediately legible. On an overcast night, temperatures often remain relatively mild. Clouds act like an infrared ceiling: they absorb radiation rising from below and emit their own infrared radiation downward as well as upward. The ground still radiates, but it receives more infrared in return. Under clear skies, by contrast, the effective radiating surface for much of the infrared window can be high, dry air and space itself. The ground radiates outward with little coming back, and it cools rapidly.

This is also why weather forecasts sometimes distinguish between the air temperature and the "feels like" temperature on clear nights. Your skin is a surface that can radiate heat to the sky. Under clear, dry conditions it can lose energy efficiently by infrared emission, making you feel colder than the air temperature alone would suggest. The exhaust is part of comfort as well as climate.

Thermal equilibrium as a living target. In a perfectly steady equilibrium, absorbed solar energy equals outgoing infrared energy. But Earth is not perfectly steady. Seasons shift where sunlight is deposited, oceans store heat and release it later, volcanic aerosols come and go, and internal variability like El Niño rearranges heat between ocean and atmosphere. The energy budget wobbles around its long-term balance. What makes the infrared exhaust crucial is that it supplies the gravitational pull back toward a well-defined relationship: if the planet accumulates extra heat, temperatures rise and infrared emission increases until the imbalance is reduced.

The ocean plays a special role here, acting as a vast heat reservoir that can absorb energy without immediate large swings in surface temperature. That reservoir smooths the response but also delays it: the planet can be out of balance for years or decades while heat is being tucked away below the surface. The infrared exhaust still governs the long-term outcome, but the approach to balance can be slow, like a heavy ship responding to a change in the engine.

The remarkable thing is that the governing mathematics remains simple enough to write on a page, even while the physical pathways are complicated enough to fill libraries. A global average absorbed solar flux of order 239 $\mathrm{W\,m^{-2}}$ corresponds, through σT^4, to an effective emission temperature near 255 K. The fourth power ensures that any sustained attempt to push Earth warmer forces a stronger and stronger radiative response. The planet does not need intention or design to have this stabilizer; it falls out of the basic physics of thermal radiation, the same physics that lets an engineer estimate the heat loss from a hot pipe, or an astronomer infer the temperature of a distant star.

On a clear night, standing under a sky full of stars, it is possible to sense the exhaust at work. The air may be still, the world quiet, and yet every warm surface—your coat, the roof tiles, the grass—is steadily sending invisible radiation upward. The stars feel cold not because they are cold, but because the space between them is so empty that it offers very little in return. Earth remains habitable because it can pay its energy bills to that emptiness, watt by watt, in the steady language of infrared light.

1.3 The Molecular Gatekeepers

The 33°C gap between a bare-rock Earth (radiating to space as though it were about 255 K) and the Earth people inhabit is not a mystery solved by adding more sunlight. The solar stipend is what it is. The gap is solved by the fact that the planet does not radiate to space from the same place it is warmed.

A useful word here is *gatekeeper*. Certain molecules in the atmosphere decide, wavelength by wavelength, whether infrared light travels freely to space or is intercepted, absorbed, and re-emitted on its way out. The main gatekeepers are familiar chemical formulas: water vapor (H_2O), carbon dioxide (CO_2), methane (CH_4), nitrous oxide (N_2O), and ozone (O_3). Clouds, though not molecules, behave like gatekeepers too, because droplets and ice crystals absorb and emit infrared over broad bands.

Calling them gatekeepers is better than calling them a "blanket." A blanket metaphor suggests a passive layer that simply keeps heat from escaping. The atmosphere is not passive, and it is not a uniform cloth. It is a three-dimensional, moving fluid with selective optical properties, threaded through by convection and weather. The crucial act is not "trapping" heat as a substance. The crucial act is changing the *altitude* and *temperature* of the place from which the planet can most efficiently radiate to space.

That place is often called the *effective emission level* or *effective emission height*. It is not a sharp surface you could fly a plane along. It is more like a statistical idea: for a given wavelength, there is a typical altitude where the atmosphere becomes thin enough that photons emitted upward have a good chance of escaping to space. When gatekeeper gases become more abundant, that typical altitude tends to rise. Higher usually means colder in the lower atmosphere, and colder means weaker infrared emission. The planet then has a problem: it is still absorbing the same sunlight, but it is emitting less infrared to space. To restore balance, temperatures below must increase until the system once again emits enough to pay its energy bills.

The greenhouse effect can be introduced without a single diagram, but it is hard to hold in the mind without a picture of vertical structure. The central physical facts are these: (i) the lower atmosphere generally gets colder with height (the *lapse rate*); (ii) infrared emission to space is set by temperature; (iii) the atmosphere is partly opaque in the infrared, so emission to space comes from different heights at different wavelengths; (iv) increasing greenhouse gases tends to raise the effective emission height at key wavelengths. Put those together and the 255 K effective radiating temperature can coexist with a warmer surface.

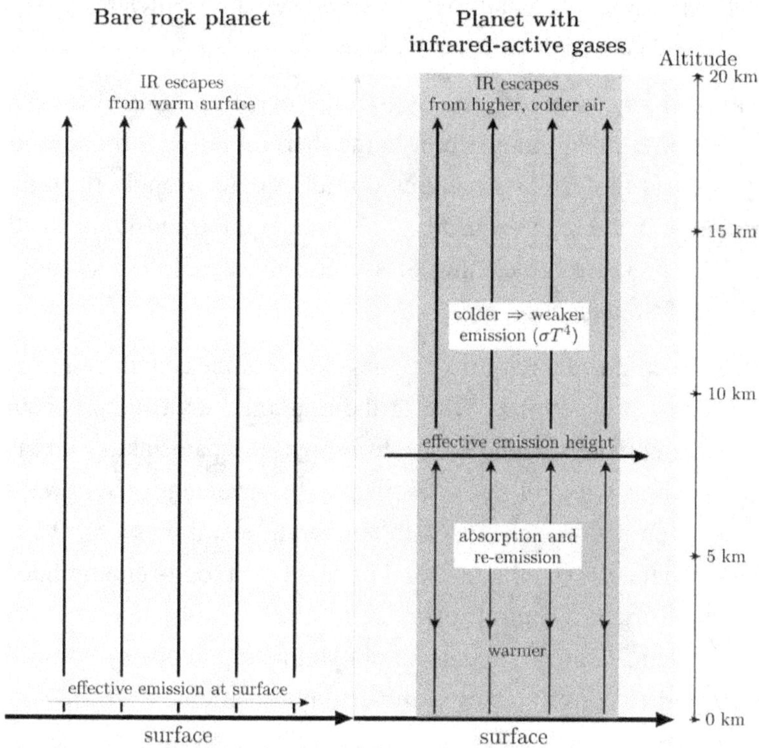

The diagram hides a subtlety that matters: the effective emission height depends on wavelength. Some infrared wavelengths escape from near the surface on clear days; others escape only from high altitudes because the air is opaque below. "Emission height" is really a family of heights, one for each slice of the spectrum, and the outgoing infrared to space is the sum of all those slices.

How molecules become gatekeepers. A greenhouse gas is not defined by moral status or by abundance. It is defined by quantum mechanics. Molecules can rotate and vibrate in specific ways, and those motions interact with electromagnetic radiation at specific wavelengths. If a molecule has vibrational modes that change its electric dipole moment, it can absorb and emit infrared radiation efficiently. Water vapor and carbon dioxide have exactly such modes.

Nitrogen (N_2) and oxygen (O_2), which make up most of the atmosphere, are largely transparent in the thermal infrared because their symmetric molecular structures do not couple strongly to infrared radiation in the relevant bands. They matter enormously for weather and for pressure broadening of spectral lines, but they are not the main infrared gatekeepers.

This spectral selectivity is not guessed. It is measured, catalogued, and constantly refined. One of the unglamorous triumphs behind modern climate science is spectroscopy: the painstaking measurement of how strongly gases absorb at each wavelength and how those absorption features change with temperature and pressure. The HITRAN database, maintained and updated by a long-running international effort, is a central reference for these line-by-line absorption parameters. Climate models do not simulate photons by consulting poetry; they consult tables of molecular transitions.

At first glance, the spectrum looks like a barcode: narrow lines where absorption is strong separated by gaps where it is weaker. In the real atmosphere, those lines broaden because molecules collide and because they move at thermal speeds, and water vapor adds a continuum-like absorption that fills in some of the gaps. That "continuum" is not an aesthetic flourish; it is one of the ways the infrared window can become less window-like in humid conditions. Radiative-transfer models often include a parameterization of this effect, such as the MT_CKD continuum model, and laboratory work continues to test and improve it because small errors in window regions can translate into meaningful differences in planetary cooling.

A nineteenth-century clue in a glass tube. The gatekeeper idea has a clean historical origin. In 1859, John Tyndall performed experiments that read today like a simple magic trick: he sent infrared radiation through tubes filled with different gases and measured how much was absorbed. Dry air was relatively transparent; water vapor and carbon dioxide absorbed strongly. Tyndall's apparatus was finicky and his measurements demanded patience because he was working near the limits of what instruments could resolve. But the implication was bold and direct: the composition of the atmosphere could affect the planet's temperature by regulating infrared radiation.

The experiment remains emotionally satisfying because it shrinks a planetary phenomenon into a tabletop demonstration. Put the right gas in the path of infrared light and the detector sees less energy. Scale that up to an atmosphere several kilometers thick and you have the first physical handle on why the Earth can have a cold effective radiating temperature and a warm surface.

Optical thickness and the "last interaction" idea. A photon of thermal infrared emitted from the surface begins a perilous journey. Along the way it may be absorbed by a greenhouse gas molecule or a cloud droplet. After absorption, the energy is typically redistributed among molecular motions and then re-emitted as a new photon in a random direction, with a spectrum characteristic of the local temperature. This is not a single bounce; it can happen many times.

In a thick enough layer, the photon's path becomes a kind of random walk. Eventually, one photon happens to be emitted upward from a sufficiently high, thin layer and escapes to space without being absorbed again. In that sense, the outgoing infrared that satellites measure is dominated by the *last* emission event before escape. The altitude of that last event is the effective emission height.

The probability of escape depends on how much absorbing material lies above, which depends on concentration, pressure, humidity, and cloudiness. Add more CO_2, and at wavelengths where CO_2 absorbs strongly, the atmosphere becomes more opaque. The last emission tends to occur higher up. If higher up is colder, the emission is weaker at those wavelengths. The outgoing infrared decreases unless temperatures rise somewhere to compensate.

This explanation is often clearer than the blanket metaphor because it identifies a mechanism that can be tested: change the opacity, the emission height moves; change the emission height, the effective radiating temperature changes; change the effective radiating temperature, the top-of-atmosphere infrared flux changes.

Why water vapor and carbon dioxide play different roles. Water vapor is the dominant greenhouse gas in the sense that it contributes a large fraction of the infrared opacity in today's atmosphere. But it is also short-lived: a parcel of moist air rains out its water on timescales of days. Water vapor concentration is therefore strongly controlled by temperature and circulation. Warm air can hold more water vapor; cold air holds less. That makes water vapor a powerful amplifier: when something else warms the planet a bit, the atmosphere tends to moisten, which increases infrared opacity and pushes the effective emission level upward, requiring further warming to restore balance.

Carbon dioxide behaves differently. It is well mixed and long lived on human timescales. It does not quickly rain out. That makes it an especially effective *control knob* for the infrared gatekeeping because its concentration can change for reasons not immediately tied to local weather, and because it persists long enough to alter the radiative balance for decades to centuries. The details of how strongly it affects outgoing infrared depend on the shape of its absorption band, pressure broadening, overlap with water vapor absorption, and the state of the atmosphere. But the conceptual role is consistent: CO_2 changes the opacity in a part of the spectrum where Earth emits a great deal of thermal radiation.

A picture helps here too, even without drawing a full spectrum. Earth's thermal emission peaks around 10 μm, near the infrared window, while CO_2 has a strong band near 15 μm. Water vapor absorbs broadly across much of the thermal infrared, including window edges and, via continuum absorption, parts of the window itself. Ozone adds an important feature near 9.6 μm. The outgoing infrared is the sum of all these wavelength-by-wavelength escape routes, each with its own gatekeepers and its own effective emission height.

Clouds: gatekeepers with two faces. Clouds complicate every simple statement because they influence both sides of the planetary budget. They reflect incoming sunlight (cooling) and absorb/emit infrared (warming), and the balance depends on cloud type, altitude, thickness, and droplet properties. High, thin cirrus clouds can be relatively transparent to sunlight but effective at trapping infrared, tending to warm. Low, bright stratus clouds reflect a lot of sunlight and can cool strongly even though they also emit infrared.

In terms of emission height, optically thick clouds behave like their own radiating surfaces. The infrared that reaches space from a cloudy column often originates near the cloud top. A high cloud top is cold, so it emits weakly, reducing outgoing infrared and warming the system beneath. A low cloud top is warmer and emits more strongly, closer to what the surface would emit, reducing the greenhouse effect in that column. Weather, in other words, moves the gatekeepers around in altitude as well as in concentration.

A compact restatement of the 33°C *puzzle.* The bare-rock calculation effectively assumed that the same temperature characterizes both the surface and the radiating level. Earth breaks that assumption. The outgoing infrared to space corresponds to a characteristic temperature around 255 K, but the surface sits closer to 288 K. The difference exists because the atmosphere makes the radiating level colder than the surface.

The simplest quantitative glimpse of this comes from the lapse rate, the typical rate at which temperature decreases with altitude in the troposphere. A common global-average figure is about $6.5\ \mathrm{K\,km^{-1}}$ (though it varies). If the characteristic emission temperature is about 33 K colder than the surface, that temperature drop corresponds to an altitude difference on the order of a few kilometers:

$$\Delta z \sim \frac{33\ \mathrm{K}}{6.5\ \mathrm{K\,km^{-1}}} \approx 5\ \mathrm{km}.$$

This is not a claim that there is a single "emission surface" exactly 5 km up. It is a sanity check: the magnitude of the greenhouse effect is consistent with the idea that much of Earth's infrared exhaust, in the aggregate, escapes from mid-tropospheric altitudes rather than directly from the ground.

That estimate also clarifies why changing greenhouse gases matters. If increased opacity raises the effective emission height by even a small fraction of a kilometer in key spectral regions, the emission temperature drops by a few kelvins. Because outgoing radiation is so sensitive to temperature, those few kelvins correspond to watts per square meter of reduced cooling to space. The planet responds by warming until the emission temperature in those regions returns to what is needed to balance absorbed sunlight.

A door that never fully closes, and a window that never fully opens. It is tempting to imagine the greenhouse effect as a door being slammed shut. In reality, the infrared spectrum is a set of partially open doors, some narrow and some wide, and the gatekeepers vary in their authority across them.

Even in a humid atmosphere, there remain spectral regions where radiation can escape more readily. Even in a dry atmosphere, there are strong absorption bands that are already quite opaque. That is why adding CO_2 does not simply "double" the greenhouse effect in a linear way: parts of its band are already saturated in the sense that the lowest layers are opaque, and additional CO_2 has its strongest incremental impact in the wings of absorption lines and in altitude regions where the opacity transition occurs. This is also why careful spectroscopy matters: the incremental changes live in the fine structure.

The window metaphor is equally imperfect. The "infrared window" is not a clean pane of glass. It is a window with smudges that thicken in humid air and with curtains that appear when clouds form. The window is one of the reasons deserts cool rapidly at night and humid tropical regions do not; it is also one of the reasons small changes in water vapor continuum absorption are not academic trivia. A slightly less transparent window changes how efficiently the surface can cool under clear skies, altering the baseline from which weather and climate operate.

A person can feel these gatekeepers at work without naming a single molecule. Step outside on a winter night beneath thick low clouds and the cold can seem muted, as though the sky itself is closer. Step outside on a clear, dry night and the cold can feel sharper, even if the air temperature is similar, because your body radiates into a sky that offers little infrared in return. The sensation is the physics made intimate: infrared leaving you, infrared not coming back, an invisible exchange shaped by the same molecular rules that govern Earth's ability to stay warm while still shining 239 $W\,m^{-2}$ of thermal radiation out into space.

Chapter 2

Engines of Amplification

While carbon dioxide pulls the trigger on climate change, it is the planet's own internal machinery that determines the caliber of the bullet. This chapter peels back the layers of the atmosphere to reveal how reactive loops can turn a modest nudge in temperature into a profound shift, or conversely, how hidden brakes attempt to slow the runaway train. We move beyond simple cause-and-effect to understand the dynamic, circular conversations taking place between heat, moisture, and radiation.

2.1 The Logic of Loops

A feedback is the simplest kind of plot twist in physics: an effect circles around and becomes a cause. Push a system, it reacts, and that reaction either leans into the push or leans against it. The words *positive* and *negative* sound like moral judgments, but they are bookkeeping. A positive feedback amplifies the original change; a negative feedback damps it. The climate problem becomes clearer, and also stranger, when you treat Earth not as a passive object being heated, but as an active, responsive machine whose responses have direction, strength, and timing.

That last word—timing—already hints at why feedbacks matter to people. A badly timed feedback is how a shower suddenly turns scalding, how a public-address system erupts into a scream, how a financial panic becomes contagious. A well-timed feedback is how your body holds a steady temperature, how a refrigerator keeps milk from spoiling, how an aircraft autopilot keeps a plane level in turbulent air. Stability is not a static condition; it is something a system *does*.

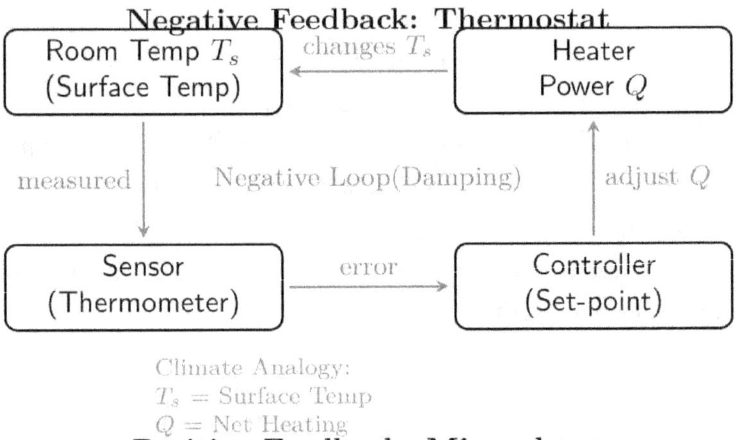

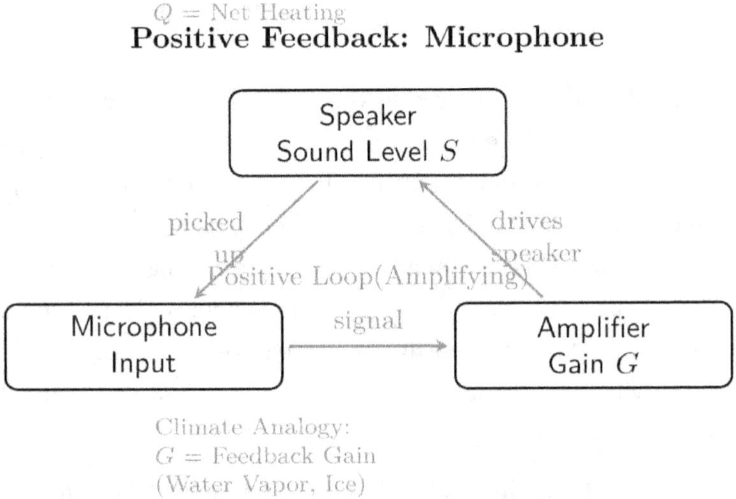

The climate system contains both kinds of loops. Some act like a thermostat: if the surface warms, the system finds ways to shed more heat to space, nudging temperatures back down. Others act like a microphone too close to a speaker: a small warming changes the atmosphere or surface in a way that makes additional warming easier. The planet's temperature, then, is not set by one knob. It is the result of competing loops, each with its own wiring.

A good way to keep the wiring straight is to start from a plain statement of energy accounting. If Earth is, on average, at a steady temperature, then the energy it absorbs from the Sun matches the energy it emits back to space as infrared radiation. A perturbation—say, an increase in carbon dioxide—upsets that balance by reducing the outgoing infrared for the same surface temperature. The planet is then gaining energy; it must warm until the outgoing infrared increases enough to restore balance. That restoration is not optional. It is demanded by conservation of energy, and it is one reason climate science can be both profoundly complicated in detail and stubbornly simple in its backbone.

The first stabilizing loop is baked into physics itself: warmer objects radiate more. For Earth, this is sometimes called the *Planck response*—not because Max Planck was thinking about climate, but because the underlying radiation law is the same. If the planet warms, it emits more infrared to space, which is a negative feedback on temperature. Think of it as the baseline brake. Without it, temperature would drift without bound. With it, warming is self-limiting in principle. The vital question becomes: how much do other feedbacks weaken that brake, or reinforce it?

The logic of feedbacks becomes far less mystical when translated into two numbers: a *forcing* and a *feedback parameter*. A forcing, usually written ΔF, is an externally imposed change in Earth's energy balance measured near the top of the atmosphere. It is a push, expressed in watts per square meter. A feedback parameter, usually written λ, tells you how the net outgoing energy changes per degree of global-mean surface warming:

$$\Delta R \approx \lambda \, \Delta T.$$

Here ΔR is the change in net radiation to space (outgoing minus incoming at the top of the atmosphere), and ΔT is the change in global mean surface temperature. If λ is negative, then warming increases net energy loss to space, and the system damps the perturbation. If λ is less negative (closer to zero), damping is weaker. If λ were positive, warming would reduce net energy loss, and you would have an amplifying runaway. Earth, fortunately, does not appear to live in that last regime for the kinds of perturbations humans are currently applying, but some components can behave locally and temporarily like it.

The balance condition at a new equilibrium is simply:

$$\Delta F + \Delta R = 0.$$

Substituting the linear relation gives:

$$\Delta F + \lambda \, \Delta T = 0 \quad \Rightarrow \quad \Delta T = -\frac{\Delta F}{\lambda}.$$

Everything in this little equation is packed with meaning. The forcing is the push. The feedback parameter is the system's tendency to push back. The minus sign is the thermostat logic: if $\lambda < 0$, a positive forcing leads to a positive temperature change until extra outgoing radiation cancels the forcing.

In practice, λ is not a single physical process. It is a sum:

$$\lambda = \lambda_0 + \lambda_{\mathrm{wv}} + \lambda_{\mathrm{lr}} + \lambda_{\mathrm{c}} + \lambda_{\alpha} + \cdots$$

where λ_0 is the Planck response (the fundamental radiative brake), and the other terms represent feedbacks from water vapour (wv), lapse rate (lr), clouds (c), and surface albedo (α), among others. The signs matter. The Planck response is strongly negative. Surface albedo feedback (ice and snow retreating, exposing darker surfaces) is positive. Water vapour feedback is positive. Lapse rate feedback tends to be negative in the tropics. Cloud feedback is the most intricate, because clouds affect both sunlight and infrared radiation and change in several ways at once.

The reason this decomposition is so useful is that it turns a messy picture into a ledger. If a process makes λ less negative, it increases warming for the same forcing. It does not need to be dramatic or intuitive to matter; it only needs to nudge the budget.

Engineers often describe this same logic using a *gain* factor. Suppose you imagine an idealized planet with only the Planck brake: you push with a forcing ΔF and the temperature change would be

$$\Delta T_0 = -\frac{\Delta F}{\lambda_0}.$$

Now let additional feedbacks respond to temperature and modify radiation. Define the combined feedback from everything except the Planck term as λ_{fb} (this includes positive and negative pieces). The actual temperature change is

$$\Delta T = -\frac{\Delta F}{\lambda_0 + \lambda_{\text{fb}}}.$$

It can be written as

$$\Delta T = \frac{\Delta T_0}{1 - g},$$

where

$$g = -\frac{\lambda_{\text{fb}}}{\lambda_0}.$$

Here g is a dimensionless gain. If g is positive, the non-Planck feedbacks oppose the Planck brake and amplify warming. If g is negative, they strengthen the brake and reduce warming. If g approaches 1 from below, the denominator shrinks and the response becomes very large: a system on the edge of instability. This is the same algebra that governs a microphone squeal. A sound system becomes unstable when the loop gain exceeds unity at some frequency; then a tiny noise gets recycled and magnified faster than losses can dissipate it.

The climate, unlike a sound system, does not have one single gain. It has many, spread across regions, seasons, and timescales. But the idea that matters for stability is the same: feedbacks act as multipliers.

That may sound abstract, so it helps to hold a real piece of history in mind. Long before climate feedbacks were quantified, the mathematics of feedback was sharpened in the smoke and clatter of the Industrial Revolution. Steam engines were powerful, but dangerous: if the engine sped up, it could overspeed into failure. The solution was the centrifugal governor, a device with spinning balls that lifted arms as speed increased, throttling the steam valve and slowing the engine back down. In 1868, James Clerk Maxwell analyzed such governors and showed that it is not enough for feedback to be negative in sign; it must also be stable in time. A governor that reacts too strongly or too slowly can make the engine hunt and oscillate. The result is a machine that lurches: speed up, clamp down, slow too much, open up, speed up again.

The climate has its own versions of hunting and overshoot. Some feedbacks are fast: water vapour changes with temperature in days. Some are slow: ice sheets and deep ocean circulation respond over centuries or longer. A fast positive feedback can create a rapid surge;

a slow negative feedback can eventually counter it, but only after large changes occur. In everyday language, this is the difference between a quick flare-up and a controlled burn. In mathematical language, it is the difference between equilibrium sensitivity (how much warming once the system settles) and transient response (how warming unfolds while parts of the system lag behind).

Even without writing down differential equations, timing changes how feedback feels in human terms. A stabilizing feedback that acts slowly is cold comfort if it arrives after coastlines have shifted. A stabilizing feedback that acts quickly is the difference between a livable heatwave and a deadly one.

The climate's feedback bookkeeping has another subtlety: feedbacks can be real and strong even when they do not *create* energy. None of these loops violate conservation laws. They work by changing the pathways through which energy flows. Water vapour changes how opaque the air is to infrared radiation. Clouds rearrange both sunlight and infrared escape routes. Ice and snow change how much sunlight is absorbed. These are not energy sources; they are valves and curtains and reflectors in the planetary heat engine.

That is why the phrase *climate sensitivity* is so central. In its simplest form, climate sensitivity is the equilibrium temperature change per unit forcing:

$$S = \frac{\Delta T}{\Delta F} = -\frac{1}{\lambda}.$$

Large sensitivity corresponds to λ being only weakly negative: the brake exists but is softened. Small sensitivity corresponds to a strongly negative λ: the brake bites hard. When scientists discuss the warming expected from a doubling of carbon dioxide, they are applying exactly this relation, with ΔF replaced by the radiative forcing associated with doubling and λ replaced by the net feedback parameter of the climate system.

There is a temptation to treat sensitivity as a single magic number, like the thermostat setting on a wall. But sensitivity is a property of the whole loop structure: the balance of fast feedbacks, slow feedbacks, and the baseline radiative brake. Change the background climate, and the loops can change strength. Move from a colder world with extensive ice to a warmer world with less, and the albedo feedback weakens. Add aerosols that affect clouds, and cloud feedbacks may shift. Warm the deep ocean over centuries, and circulation patterns change, subtly altering where heat is stored and released. Sensitivity is not a decree; it is the current personality of a complicated machine.

Still, the linear feedback framework earns its keep because it is not merely a metaphor. It is the first-order Taylor expansion of Earth's radiative balance around its current state. Small-to-moderate changes can often be usefully approximated as linear in global mean temperature, even though the underlying physics is richly nonlinear. The expansion does not claim that the world is linear; it claims that if you do not shove too far, the tangent line is a good guide. Engineers use the same trick when they linearize a system around an operating point to analyze stability. It is how you take an aircraft that is complicated in full motion and still ask, in a crisp way, whether it tends to return to level flight after a gust.

A concrete climate example makes the abstraction tangible. Consider sea ice in the Arctic at the end of summer. Ice is bright; open ocean is dark. If a warm summer melts more ice than usual, more dark water is exposed. Dark water absorbs more sunlight than ice, which warms the ocean surface, which melts more ice. That is a positive feedback loop, and it helps explain why Arctic warming has been so rapid compared with the global average. Yet the loop does not run away without limit. Winter returns, the Sun disappears for months, and the ocean begins losing heat to the cold air above it. Ice reforms. There are stabilizing processes and seasonal resets. Positive feedback does not mean inevitable catastrophe; it means the push is being helped rather than resisted.

Or take a negative feedback that is so familiar it is easy to overlook: the way a hot day can build towering thunderstorms. Warm air holds more water vapour and fuels stronger convection; rising air forms clouds that reflect sunlight and can cool the surface beneath. That looks like a thermostat. But the same convective clouds can also trap infrared radiation, especially at night, and their net effect depends on cloud type, height, and timing. The sign of the cloud feedback is not written on the cloud itself. It emerges from the loop: surface warming changes cloud properties; cloud properties change radiation; radiation changes surface warming. Getting the sign and magnitude right requires care, measurement, and models that respect the physics.

The most important lesson of loop logic is that intuition is an unreliable guide to strength. A feedback can feel minor and still matter if it nudges λ toward zero. Suppose the Planck brake alone would give a certain warming, and the net effect of all other feedbacks is to cut the magnitude of λ in half. Then the warming doubles. No single feedback needs to be dominant for the total amplification to be large; the system only cares about the sum in the denominator.

Loop logic also guards against a common misunderstanding: that feedbacks are optional add-ons to a greenhouse-gas calculation. They are not decorations; they are the climate system responding. If carbon dioxide warms the surface even a little, the atmosphere and surface will not sit still. Water will evaporate differently. Ice and snow will shift. Clouds will reorganize. The only question is how those responses add up in radiation space.

It is worth lingering on one more idea that makes feedback analysis feel real rather than symbolic: the distinction between *forcing* and *feedback* is partly a choice about what you treat as external. Carbon dioxide emitted by humans is usually treated as an external forcing. Water vapour is treated as a feedback because its atmospheric concentration is mainly set by temperature through evaporation and condensation. But nature does not label variables in advance. Over long enough timescales, carbon dioxide itself can behave like a feedback because temperature alters the carbon cycle: oceans absorb CO_2 differently as they warm, ecosystems change, and weathering rates shift. The same physical mechanism can move categories depending on the time window. That is not a flaw in the concept; it is a reminder that a loop diagram is a model of a particular question asked over a particular span of time.

If all of this seems like a lot of machinery for what amounts to "how much will it warm?," it helps to remember why the machinery exists. A planet cannot negotiate with the Sun, and it cannot vote on energy conservation. It can only redistribute energy flows until the books balance. Feedbacks are the rules of redistribution. They are the difference between a small shove and a large response, between a gentle adjustment and a sharp lurch. And because the strongest feedbacks act through ordinary substances—water vapour, clouds, ice, and the vertical temperature structure of air—the planet's stability depends on the everyday physics of evaporation, condensation, reflection, and radiation.

A thermostat on a wall is reassuring because it makes stability visible: a number you can set, a click you can hear. Earth's thermostat is quieter. Its loops run through the transparency of air and the brightness of clouds, through a few percent changes in humidity high above your head, through the timing of melt and freeze at the margins of continents. The mathematics does not replace wonder; it simply insists that wonder has accounting rules, and that amplification and damping are not moods of nature but properties that can be measured, compared, and added up.

2.2 The Humidity Multiplier

Carbon dioxide often gets cast as the main character in the greenhouse story, but the atmosphere's most abundant greenhouse gas is water vapour. That simple fact can mislead in two opposite directions. One mistake is to say: if water vapour dominates, why worry about CO_2? The other is to say: if CO_2 is the control knob, then water vapour is merely background noise. The more accurate picture is closer to how kindling and logs work in a fire. Carbon dioxide is long-lived and accumulates; it provides the persistent push. Water vapour responds quickly; it provides a powerful echo.

The echo begins with a piece of everyday experience that hides a law of thermodynamics. On a cold winter day, your breath appears as a little cloud. On a hot summer day, the same breath is invisible. The difference is not your lungs but the air. Cold air cannot hold much water vapour before the excess condenses into droplets. Warm air can hold much more in gaseous form. This is not a vague tendency. It is a quantitative constraint set by the balance between molecules escaping from liquid water and molecules returning to it, a balance that shifts strongly with temperature.

That constraint is usually introduced with the Clausius—Clapeyron relation, which links the saturation vapour pressure (the partial pressure of water vapour at which condensation begins) to temperature. The name sounds forbidding, but its message is intuitive: warming makes evaporation easier because more molecules have enough energy to leave the liquid surface, and it makes condensation harder because the vapour needs a higher pressure to be forced back into liquid. When the math is worked out for Earth-like temperatures, the saturation vapour pressure rises by roughly 7% per degree Celsius. This number is not a guess and not a fit to modern data; it comes from the latent heat of vaporization of water and the gas constants. It is the planet-sized reason that humidity matters.

Humidity itself needs a small vocabulary lesson. *Absolute humidity* means the actual amount of water vapour in the air, often expressed as a mixing ratio (grams of water per kilogram of air) or as a vapour pressure. *Relative humidity* is a ratio: how close the air is to saturation at that temperature. Warm air can have a high absolute humidity while having a modest relative humidity; cold air can have a low absolute humidity while still being near saturation. Weather reports emphasize relative humidity because it connects to comfort and to condensation, but radiative physics often cares more about absolute humidity because it determines how strongly the atmosphere absorbs infrared radiation.

The humidity multiplier emerges from an empirical regularity: on large scales, relative humidity tends to remain roughly similar as climate warms. Not constant everywhere and not constant every day—storms, deserts, and ocean boundaries all have their own quirks—but stable enough in the global mean for a first-order statement: as temperature rises, the atmosphere tends to take on more water vapour in proportion to its rising capacity. That means absolute humidity increases with warming, and because water vapour absorbs infrared radiation, the greenhouse effect strengthens. A little warming invites more water vapour; more water vapour traps more heat; more trapped heat causes more warming.

It is worth pausing over why relative humidity might stay roughly similar. The atmosphere is not a sealed jar; it is a restless engine powered by sunlight and cooled from above by radiation to space. Evaporation from the ocean surface loads the lower atmosphere with moisture. Air rises, expands, and cools; water vapour condenses into cloud droplets, releasing latent heat that helps power convection. The condensed water falls out as rain or snow, leaving the upper air much drier than the air below. This continual cycling sets a kind of statistical humidity structure that is controlled by circulation and condensation thresholds. When the system warms, evaporation and condensation both intensify. The overall pattern shifts, but the physics that prevents air from staying far above saturation still operates. The result is that the atmosphere tends to track saturation, not wander arbitrarily away from it.

The radiative side of the story hinges on what water vapour does to infrared light. The surface and lower atmosphere emit longwave radiation. Some wavelengths slip through relatively freely; others are absorbed strongly by greenhouse gases. Water vapour has a dense forest of absorption lines across much of the infrared spectrum. Add water vapour and you thicken the blanket, but not evenly. The extra opacity matters most where the atmosphere was previously semi-transparent. In those bands, the effective level from which radiation escapes to space shifts upward to higher, colder air. Colder air emits less infrared, so the outgoing radiation decreases for the same surface temperature. Energy accumulates until warming restores the balance.

This is why the water vapour feedback is counted among the "fast" feedbacks: it responds on timescales of days to weeks, limited by evaporation, condensation, and atmospheric transport, not by geological cycles. You can watch the ingredients in real time. When the ocean surface warms during an El Niño event, atmospheric water vapour increases measurably. Satellites track the column water vapour; radiosondes measure humidity profiles; reanalyses stitch it into a global picture. The atmosphere is not waiting for ice sheets or continents to shift. It reacts almost immediately.

A concrete example makes the multiplier feel less like an abstract loop and more like a physical weight on the scales. Consider a heatwave in a coastal city. Two versions of the heatwave are possible. In the first, the air is hot but dry, and nights cool down because the atmosphere is relatively transparent in parts of the infrared. In the second, moisture surges in from a warm ocean or a nearby wetland region. The day becomes oppressive because sweat evaporates less efficiently. The night stays warmer because water vapour traps infrared radiation, reducing the rate at which the surface can shed heat. People experience this difference directly: the "sticky" heat that never quite releases its grip. The same physics scales up globally. A warmer world is, on average, a moister world, and a moister world holds on to heat more stubbornly.

The amplification can be expressed in the language of feedback parameters introduced earlier. Imagine, as a mental baseline, that the planet warms in response to CO_2 forcing while holding absolute humidity fixed. In that hypothetical world, warming still increases outgoing radiation through the Planck response, and an equilibrium is reached. Now allow water vapour to increase with temperature in the real world. The increase in humidity reduces outgoing longwave radiation compared with the fixed-humidity case, effectively weakening the stabilizing radiative brake. In the bookkeeping, that appears as a positive feedback term: λ_{wv} is positive in the sign convention where negative values represent damping. The net λ becomes less negative, so the same CO_2 forcing produces a larger ΔT.

The surprising part is not that the feedback exists, but how potent it is. Water vapour is already abundant, so one might guess that adding a bit more would not matter much, as though a blanket already thick cannot be thickened further. Radiative transfer does not work that way. Because absorption depends on wavelength, pressure, and the vertical distribution of vapour, additional moisture continues to matter, especially in the upper troposphere where the air is cold and thin and where small absolute changes can be large relative changes. Held and Soden (2000) emphasized this leverage: changes aloft can disproportionately alter the outgoing infrared because the radiation escaping to space is sensitive to humidity in those layers. In effect, the upper troposphere acts like a critical valve in the infrared plumbing.

This is also the reason water vapour feedback is tied to circulation. The upper troposphere is not humid because it is close to the ocean surface; it is humid because air has been carried upward in convective regions and then spread out. Shifts in tropical convection, storm tracks, and the width of the Hadley circulation change where moisture is lofted and where it dries out. The feedback is therefore both thermodynamic (warm air holds more moisture) and dynamic (circulation decides where that moisture goes). The thermodynamics sets the broad constraint; dynamics draws the map.

A historical anecdote shows how early scientists stumbled onto this multiplier. In the 1960s, Syukuro Manabe and Richard Wetherald built one of the first radiative—convective models capable of asking a then-brazen question: what happens to temperature if CO_2 increases? Their 1967 work compared different ways of treating humidity. If absolute humidity were held fixed as temperature changed, the warming was more modest. If relative humidity were specified in a way that allowed absolute humidity to increase with warming, the sensitivity rose sharply. The result was not a detail. It was a demonstration that the atmosphere's water cycle is not a side show; it is part of the amplifier that turns a forcing into a larger temperature response. The computational tools of the time were crude by modern standards, but the conceptual punch landed: humidity assumptions change the answer.

The water vapour feedback, however, does not act alone. It has a frequent partner that behaves like a partial counterweight: the lapse-rate feedback. The lapse rate is the rate at which temperature decreases with height in the troposphere. In humid, convective regions, the vertical temperature profile is constrained by moist convection, which tends to warm the upper troposphere more than the surface as the climate warms. That pattern increases the efficiency with which the atmosphere can radiate to space, contributing a negative feedback on surface temperature in the tropics. Models and theory show that water vapour and lapse-rate changes are tightly coupled: the same moist convection that lofts water also shapes the vertical warming pattern. Soden and Held (2006) found that while lapse-rate feedback varies more among models than water vapour feedback does, the sum of the two has relatively small spread because they compensate each other to some degree. In plain language, the humidity amplifier and the atmospheric "radiator profile" are linked gears in the same machine.

This coupling is helpful for understanding why the humidity multiplier does not imply runaway warming. A positive feedback does not mean the climate "feeds on itself" without limits. It means that part of the response helps the initial change. The system still contains strong stabilizing physics: warmer temperatures increase infrared emission overall; the vertical temperature structure changes; clouds and circulation patterns respond in ways that can either amplify or damp. The combined effect is that the total feedback parameter remains negative, so a new equilibrium is reached. The humidity multiplier makes that equilibrium warmer than it would otherwise be.

The multiplier also clarifies why discussions of "saturation" can be misleading. People sometimes picture the greenhouse effect as a glass that can fill up. Once full, more CO_2 or more water vapour would do nothing. Real gases do saturate *at specific wavelengths* and *at specific pressures*, but the atmosphere is a layered medium with pressure broadening of absorption lines near the surface and different emitting levels across the spectrum. Adding greenhouse gases continues to matter by shifting emission levels upward and by affecting the wings of absorption bands and spectral regions that were not fully opaque. Water vapour, with its wide spectral reach, remains influential even in a humid climate.

There is also a human-scale implication that feels almost unfair. Water vapour is not something societies can directly regulate in the same way as CO_2 emissions. You cannot legislate drier air. Water vapour rains out in days; the atmosphere forgets it quickly. That would be a blessing if water vapour were the main driver. It is a complication because water vapour is a powerful responder: it takes the warming set in motion by long-lived gases and intensifies it. This is why focusing on CO_2 is not a refusal to acknowledge water vapour; it is a recognition of control. Altering CO_2 changes temperature, and temperature changes water vapour.

The feedback can be felt not just in temperature but in the hydrological cycle's intensification. More water vapour means more available fuel for heavy precipitation. The same Clausius—Clapeyron scaling that strengthens the greenhouse effect also raises the ceiling on downpours. A warmer atmosphere can carry more moisture into storms, and when storms release that moisture, rainfall rates can increase. The humidity multiplier is thus not only an infrared story but a societal one: it links warming to floods, infrastructure stress, crop damage, and public health. A few percent change per degree sounds gentle until it is applied to a system that already moves enormous quantities of water.

A subtle point matters here. The atmosphere can hold more water vapour, but it does not automatically become uniformly wetter. Many regions are projected to get drier at the surface even as the global mean water vapour increases. The reason is circulation: warming tends to intensify existing patterns of moisture convergence and divergence. Wet regions often become wetter (more moisture converges into storms), while subtropical dry zones can expand or intensify as descending air warms and dries. The humidity multiplier is global in the sense of total water vapour, but its lived experience is regional.

Even the everyday word "humidity" can be confusing because people often associate it with foggy mornings and sticky afternoons, not with dry high altitudes. Yet for radiation, the thin, cold air several kilometres up can matter disproportionately. If a small warming increases specific humidity aloft by a modest amount, the fractional change can be large because the baseline is so dry. And because infrared emission to space emerges from these higher layers in many bands, the radiative impact can be substantial. It is one of those climate facts that feels counterintuitive at first: a gram of water vapour near the tropopause can matter more for outgoing radiation than a gram near the surface, even though the surface is where most water resides. The greenhouse effect is not a blanket lying on the ground; it is a three-dimensional filter through which radiation must climb.

For a moment, return to the simple gain idea. A forcing from CO_2 produces an initial warming. The humidity multiplier increases the warming by feeding back on infrared escape. The magnitude of the amplification depends on how strongly humidity increases with temperature (the thermodynamic scaling), how humidity is distributed vertically and horizontally (the dynamic response), and how that humidity interacts with radiation (the spectral physics). Those ingredients are complex, but they do not float free. They are constrained by conservation laws and by measurable quantities like relative humidity, precipitation, and outgoing longwave radiation. That is why the water vapour feedback, among all feedbacks, is often described as one of the most robust: it is anchored by a strong thermodynamic relationship and observed patterns of humidity.

The most vivid way to appreciate that robustness is to imagine the opposite: a warming world in which the atmosphere somehow became drier in absolute terms. Such a world would fight warming by opening infrared escape routes; it would behave like an implausible self-cooling mechanism. To get there, evaporation would need to weaken or condensation would need to remove moisture so efficiently that the atmosphere loses vapour even as saturation rises. That would require a reorganized circulation that persistently wrings out the air more than today while maintaining the same energy flows—a tall order given the tight coupling between latent heat release and atmospheric motion. Nature does not forbid regional drying, but global drying in the face of warming would be a thermodynamic rebellion.

There is also a psychological aspect to the humidity multiplier. People often think of water as a moderator: oceans buffer temperature swings, and evaporation cools the skin. Both are true in the right context. Yet water vapour in the air is also an amplifier of heat retention. The same molecule that cools you through evaporation when it leaves your skin warms you when it stays airborne and absorbs infrared radiation. The direction depends on where in the energy budget you are standing. Climate feedbacks are full of such dual roles, and water is the most dramatic example because it changes phase, carries latent heat, and interacts strongly with radiation.

A final image can anchor the idea without pretending the world is simple. Picture the planet at night from space, not in visible light but in infrared. The brightness patterns tell you about temperature and about how transparent the atmosphere is. In dry regions under clear skies, infrared escapes more directly from lower, warmer levels. In humid tropical regions, much of the infrared you see comes from higher, colder layers because water vapour and clouds have shifted the effective emitting level upward. Now imagine turning the global thermostat up by a degree. The dry places brighten a bit because warmer surfaces emit more. The humid places do not brighten as much, because additional water vapour makes the atmosphere more opaque and keeps the effective emitting level high and cold. The planet's ability to dump heat to space is reduced precisely where the air is wettest, and that reduction is one of the ways a small warming becomes a larger one.

Water vapour is sometimes called a "feedback" as though that were a technical footnote. In practice it behaves like a multiplier that sits close to the heart of the greenhouse effect. It does not initiate the warming, but once warming begins, it changes the rules of infrared escape quickly and powerfully, turning temperature into humidity and humidity back into temperature, until the energy books balance again at a higher line.

2.3 The Radiator in the Sky

A stove is easiest to judge when you can feel the heat on your hands. Earth's heat is judged from far away, by space. The planet cools itself mainly by emitting infrared radiation upward, and much of that radiation does not come straight from the ground. It comes from the atmosphere, and not evenly from all heights. Some wavelengths escape from low, warm air; others escape only after climbing to high, cold layers where the air is thin enough to let photons slip through. In that sense, the climate's "radiator" is not a metal grille you can touch. It is a set of invisible emitting surfaces stacked in the sky, each one defined by opacity.

This verticality matters because warming is not uniform with altitude. The surface can warm by a degree while the upper troposphere warms by more, or less, depending on where you are and how convection reorganizes. Those vertical patterns feed back on radiation. The lapse rate feedback is the name given to the piece of the climate response that comes from changes in the vertical temperature gradient: the lapse rate itself. It is one of the most counterintuitive stabilizers in the system, because it often works by warming the upper atmosphere *more* than the surface in the tropics. A warmer upper troposphere sounds, at first, like a thicker blanket. In practice it can act like opening the radiator valve.

The lapse rate is simply the rate at which temperature decreases with height, often written $\Gamma = -dT/dz$. In the troposphere, temperature generally falls with altitude because air that rises expands in lower pressure and cools. If the rising air is dry, it cools quickly—about 9.8°C per kilometre, the dry adiabatic lapse rate. If the rising air is moist, water vapour condenses as it cools, releasing latent heat that offsets some of the cooling. The moist adiabatic lapse rate is therefore smaller in magnitude and depends on temperature: in warm, humid air it can be closer to 4°C to 6°C per kilometre; in cold air it approaches the dry value. The atmosphere is forever negotiating between these two limits, and the tropics, with abundant moisture and deep convection, are where the moist constraint most strongly shapes the vertical profile.

Now add warming. If the lower troposphere warms, the saturation vapour pressure rises, moist convection carries more latent heat, and the moist adiabatic lapse rate tends to decrease in magnitude: the temperature declines more gently with height. A gentler decline means that, compared with the surface, the upper troposphere becomes warmer than it would under a uniform warming with height. This is why many models show a "hot spot" in the tropical upper troposphere when the surface warms, regardless of the cause of warming. It is not a fingerprint of a particular forcing; it is a fingerprint of moist convection and thermodynamics.

Why does this matter for radiation? Because the outgoing infrared flux depends strongly on the temperature of the layers from which radiation escapes to space. If the upper troposphere warms, those emitting layers warm. Warmer emitting layers radiate more effectively, increasing outgoing longwave radiation and therefore cooling the system. In the feedback bookkeeping, this is a negative feedback on surface temperature: it strengthens the radiative loss for a given surface warming, nudging the net feedback parameter λ to be more negative. In other words, it helps the Planck brake bite.

The "radiator in the sky" is therefore not merely a poetic image. It is literally a layer of air whose temperature has a large leverage over how much infrared escapes. Many wavelengths in the thermal spectrum are shaped by water vapour absorption; photons in those bands may last interact with the atmosphere somewhere in the mid-to-upper troposphere before heading to space. If those layers warm faster than the surface, the planet can rid itself of heat more efficiently than it would if the warming were locked near the ground.

At this point, it is tempting to ask: if upper-air warming helps cooling, why worry about water vapour at all? The answer is that water vapour and lapse rate responses are coupled in a tug-of-war. More water vapour tends to reduce outgoing longwave radiation (a positive feedback), while a reduced lapse rate in the tropics tends to increase outgoing longwave radiation (a negative feedback). The same moist processes that load the atmosphere with vapour also reshape the vertical temperature profile. One reason climate sensitivity is not wildly uncertain is that these two feedbacks partly offset, like two people on opposite ends of a rope, each pulling harder as the other does. Soden and Held (2006) highlighted that while lapse rate feedback varies more across models than water vapour feedback, their sum tends to be comparatively stable because the physics links them.

This coupling can be felt in the rhythm of tropical weather. A hot, humid day over the ocean often ends with towering cumulonimbus clouds. Those clouds are the visible product of moist convection: warm, buoyant air rising, condensing, and releasing latent heat. Most of that latent heat is not left at the surface. It is deposited aloft, warming the mid and upper troposphere. The storm is not only redistributing water; it is redistributing heat vertically. That vertical redistribution is a local, short-lived cousin of the global lapse rate response. Scale up from one storm to millions, and the atmosphere's tendency to move heat upward becomes a key part of the global energy budget.

A concrete experience makes the radiative consequences easier to imagine. Consider the difference between a clear, dry desert night and a humid tropical night. In the desert, the ground cools rapidly after sunset. The air near the surface chills, sometimes producing a strong temperature inversion: warmer air above cooler air. The sky feels like a vast cold sink. In the humid tropics, nights cool more slowly. Part of that is clouds and water vapour trapping infrared radiation, but part is also that the temperature profile aloft is different: convection and latent heat release keep the upper troposphere relatively warm, and the emitting layers are not as cold a sink. The vertical profile is not an abstract line on a meteorological chart; it is the reason some places lose heat easily and others hold it.

The lapse rate feedback also helps explain why warming differs between regions. In the tropics, deep convection strongly couples surface temperatures to the temperature aloft, enforcing something close to a moist-adiabatic structure. This makes the upper troposphere warm relatively quickly, enhancing radiative cooling and providing a stabilizing influence on surface warming. At higher latitudes, convection is often shallower and moisture is scarcer; the lapse rate response can look different, sometimes even changing sign in terms of its feedback effect. The same term "lapse rate feedback" covers a global mean, but the underlying physics plays out unevenly over the planet.

To see the logic more sharply, it helps to picture the atmosphere as a set of semi-transparent layers. For each wavelength, there is an effective emission level—the altitude from which most of the radiation escaping to space at that wavelength is last emitted. Increase greenhouse gas concentrations, and those effective levels move upward to colder air, reducing outgoing radiation. Warm the atmosphere, and those levels may not move much, but their temperatures rise, increasing outgoing radiation. Water vapour changes both the opacity (moving emission levels) and, indirectly through moist convection, the temperature profile (changing the temperatures at those levels). The lapse rate feedback is the part of the story that comes from temperature structure, not from opacity.

One way to anchor this in history is to recall that the vertical structure of the atmosphere was not always a given. Early meteorology was largely a surface science: pressure readings, winds, temperature at ground stations. The realization that the atmosphere has layers with distinct behaviours came with balloon soundings in the late nineteenth and early twentieth centuries. Léon Teisserenc de Bort, launching unmanned balloons in the 1890s, found that temperature stops falling with height above a certain level and begins to level off: the stratosphere. That discovery was not directly about climate, but it mattered for climate thinking because it established that radiation and dynamics create a layered atmosphere. The troposphere is where weather happens and where convection mixes heat and moisture; the stratosphere is more stable and responds differently to changes in greenhouse gases. Once you accept that the sky has architecture, you are forced to accept that warming can have shape, not just magnitude.

The lapse rate feedback is mainly a tropospheric story, but the stratosphere hovers nearby as a reminder that radiative physics has vertical consequences. Adding CO_2 tends to cool the stratosphere because, in the thin dry air above most water vapour, CO_2 is an efficient radiator to space. That stratospheric cooling is a separate response often grouped under "adjustments" rather than surface temperature feedbacks, but it underlines a general principle: whether warming or cooling occurs in a layer depends on how that layer participates in the radiative exchange with space. The atmosphere is not warmed uniformly like a pot on a stove. Each layer has its own radiative job.

The phrase "radiator in the sky" can also mislead if taken too literally, because clouds complicate everything. High clouds are cold, and cold objects emit less infrared. A high cloud deck can therefore reduce outgoing longwave radiation compared with clear skies, acting like an extra blanket. Yet high clouds form within the same convective systems that shape the lapse rate. They respond to warming in their own ways, including changes in altitude, thickness, and coverage. Hartmann and Larson (2002) proposed an influential idea about tropical anvil clouds: there may be a preferred temperature level near which deep convective anvils detrain, related to the inefficiency of radiative cooling at very cold temperatures. If anvil tops tend to remain near a fixed emission temperature as the surface warms, then their longwave effect becomes amplifying: the surface warms, but the cloud-top emission does not increase as much, reducing the increase in outgoing longwave radiation that would otherwise cool the planet. Zelinka and Hartmann (2010) argued that a tendency for high clouds to rise with warming leads to a positive longwave cloud feedback in many models, for similar reasons.

These cloud ideas sit alongside lapse rate feedback rather than replacing it, and the distinction matters. Lapse rate feedback describes how the *temperature profile* changes under largely clear-sky physics, while cloud feedback describes how *cloud properties* change and alter radiation. The two are intertwined in reality, because convection shapes both temperature structure and cloud formation, but separating them helps keep the accounting honest. One can have a stabilizing lapse rate feedback and an amplifying cloud feedback operating in the same region, like two hands turning different knobs on the same radio.

There is a more intimate, almost domestic way to understand why the vertical profile is such a powerful lever. Think of an attic in summer. If the attic is poorly ventilated, heat accumulates under the roof, and the living space below becomes harder to cool. If ventilation improves, the attic can get hot while the rooms stay more comfortable, because the hot air is being allowed to dump heat outward. The climate system's "ventilation" is radiation to space. Where the effective emitting layers are, and how warm they are, determines how easily the planet cools. A warming upper troposphere in the tropics, by raising the temperature of those emitting layers, can function like better ventilation, increasing the outward heat flow. It is not comforting in itself to know that the upper air is warmer; the comfort is in what that warmth does to the balance of radiation.

A misconception worth disarming is the idea that a warmer upper atmosphere must always mean a stronger greenhouse effect. The greenhouse effect is not simply "heat trapped by warm air." It is a radiative phenomenon governed by opacity and emission temperatures. Warm air can be a better emitter. If a layer is the one doing much of the emitting to space in certain wavelengths, warming that layer increases outgoing radiation. In that case, upper-air warming is cooling for the surface, not warming. The sign depends on which part of the radiative pathway you are perturbing: opacity pushes emission upward and cooler; temperature pushes emission warmer where it occurs.

One can even phrase the lapse rate feedback in terms that echo everyday intuition. The greenhouse effect is often described as raising the effective height from which Earth radiates to space: more greenhouse gases mean radiation escapes from higher, colder air. To compensate, the surface must warm so that those higher layers also warm enough to radiate the required energy. The lapse rate feedback describes how efficiently surface warming is communicated upward. If the upper troposphere warms strongly when the surface warms, then the emitting layers warm efficiently, and less surface warming is needed to restore balance. That is stabilizing. If the upper troposphere warms weakly relative to the surface, then the emitting layers lag behind, and the surface must warm more to push enough heat upward into the radiating layers. That is less stabilizing.

In the tropics, moist convection makes the upward communication efficient. In drier regions, it can be less so. This is one reason the lapse rate feedback tends to be negative overall (stabilizing) in global averages: the tropics cover a large area, and their convective engine is powerful. But it is also why the global mean hides variation: the same planet can have stabilizing behaviour in one region and weaker stabilization in another, with the global sum reflecting the competition.

The lapse rate feedback also touches on a humanly resonant theme: the importance of height. People live at the bottom of the atmosphere, but the planet's cooling is judged from the top. Actions at the surface—burning fuel, changing land cover, emitting aerosols—alter an energy budget that must ultimately be satisfied by radiation from layers kilometres above our heads. The vertical structure of air, shaped by storms and condensation and the slow pulse of rising and sinking motion, becomes part of the answer to a question asked in parliament buildings and courtrooms: how much warming follows from a given push?

When atmospheric scientists plot temperature with height, the curves can look like private scribbles of a technical guild. Yet those curves encode whether the radiator is running efficiently. A gentler lapse rate in the tropics, meaning a relatively warmer upper troposphere for a given surface warming, is one of the ways the climate system resists change. It is a resistance that does not announce itself with a dramatic threshold. It is a quiet shift in where heat is placed.

A world without that shift would be stranger than it sounds. If the surface warmed but the upper troposphere did not, the atmosphere would become more strongly stratified, convection would be suppressed, and the very machinery that moves heat upward would weaken. The climate would then rely more heavily on slower processes to adjust, and surface extremes could become sharper. The lapse rate feedback is therefore not only about the mean temperature. It is about the atmosphere's habit of moving energy away from where we live and toward where it can escape.

At dusk in the tropics, when the tallest clouds turn grey and their anvils stretch like brushstrokes across the sky, it is easy to feel that the atmosphere is mostly water and motion. The lapse rate feedback is the mathematical shadow of that feeling: an acknowledgement that the planet's stability depends on the vertical choreography of moist air, and that some of the most effective cooling happens not by changing the surface directly, but by warming the right parts of the sky so that Earth can radiate its way back toward balance.

Chapter 3

A World of White and Blue

Viewed from space, our planet is a patchwork of deep, heat-hungry blues and blinding, reflective whites. This chapter investigates how these contrasting surfaces maintain our climate's equilibrium and why the shrinking of Earth's icy mirror triggers a self-reinforcing loop of warming that extends far beyond the poles.

3.1 The Planetary Mirror

Stand outside on a clear winter day and you can feel a small piece of planetary physics on your face. The sun may be low, the air may be bitter, yet the light can still seem sharp enough to squint at, bouncing off snow in a way that makes the world glitter and your eyes water. Now picture the same sunlight falling on open water: the surface looks darker, calmer, almost like it is swallowing the light rather than throwing it back. That contrast is not just aesthetic. It is an energy transaction, and the receipt is kept in temperature.

The name for the planet's reflectivity is *albedo*, from a Latin root meaning "whiteness." Albedo is simply the fraction of incoming sunlight that a surface reflects back upward rather than absorbing. An albedo of 0 means perfect absorption: every photon that arrives stays. An albedo of 1 means perfect reflection: nothing is taken in. Real surfaces live between these extremes, but not modestly between them. Some are so reflective that they behave like a mirror made of crystals; others are so absorbent that they behave like a black cloth laid under a heat lamp. The planet is quilted together from both kinds, and the arrangement of that quilt is one of Earth's first lines of defense against solar heating.

Sunlight arrives mostly as shortwave radiation: visible light and its near neighbors in the spectrum. After the surface absorbs some of that energy, it must get rid of it again, largely by emitting longwave (infrared) radiation back toward space. The familiar story of the greenhouse effect concerns how that outgoing infrared is slowed down by gases in the atmosphere. The albedo story sits earlier in the chain: it is the gatekeeper deciding how much sunlight gets past the door in the first place. A small change in that gatekeeping can matter because it acts on the whole enormous stream of solar energy that falls on the top of the atmosphere every day.

One reason albedo is so gripping as an idea is that it turns the color palette of the world into arithmetic. Fresh snow can reflect on the order of 80–90% of incoming sunlight under many conditions. Open ocean, by contrast, can reflect less than 10% when the sun is high, absorbing the rest. Sea ice sits between: often reflective, but less so than fresh snow, and much less so when it is old, wet, dirty, or riddled with melt ponds. Deserts can be bright; forests tend to be darker than we intuit, especially from the viewpoint of space. Clouds complicate everything further: they can be among the most reflective features on Earth, and they move, grow, shrink, and change character from hour to hour. Yet the basic picture is robust enough to keep in your head at a dinner table: *white surfaces tend to send sunlight away; dark surfaces tend to keep it.*

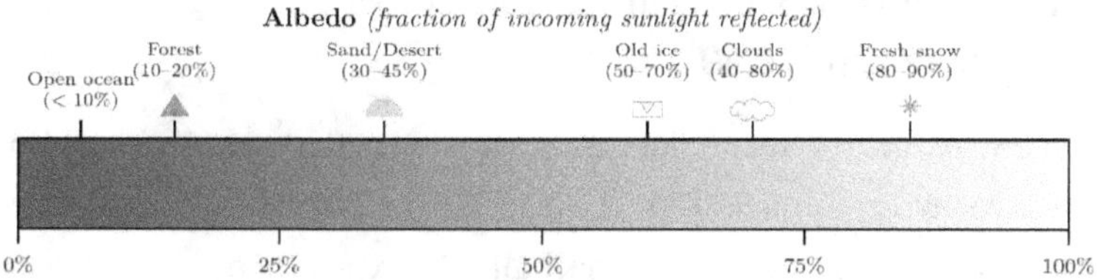

Representative albedo values. Exact reflectivity varies with sun angle, wetness, grain size, impurities, and cloud type, but the contrast between dark ocean and bright snow is consistently extreme.

A first surprise is that albedo is not a fixed property like the atomic weight of carbon. It is more like a mood, sensitive to context. The same patch of snow reflects differently at noon than at dusk, because the angle of the sun changes how much light is trapped between grains and how much skips away. Newly fallen snow is a forest of tiny prisms; over time the grains grow, edges round off, and the surface can darken. Add soot from a wildfire or the exhaust of a city, and you have sprinkled a handful of tiny heaters across a blanket that was designed to stay cool. Sea ice can be dazzling when it is dry and snow-covered; it can also turn into a patchwork of blue ponds in summer, each pond acting like a small copy of the ocean embedded in ice. Even the ocean can look mirror-like when the sun is low, reflecting a bright path toward the horizon; but the key detail is that such reflection is mostly specular, like a mirror, sending light in a particular direction. From the sun's point of view, and from the point of view of the planet's overall heat budget, open water still tends to take in a great deal of energy.

The planet's average albedo is about 0.30: roughly thirty percent of incoming sunlight is reflected back to space by a combination of clouds, atmospheric particles, and the surface below. That number is deceptively calm. It is an average over deserts and oceans, over bright cloud decks and dark forests, over polar day and polar night, over weather systems that live for days and ice sheets that live for millennia. It is also, crucially, a number that can shift. Earth is not obliged to remain a 0.30 planet.

To see why this matters, take a concrete, everyday comparison. If you have ever worn a black T-shirt and a white T-shirt on equally sunny days, you have performed a small albedo experiment with your own skin as the thermometer. The black fabric absorbs more sunlight and warms faster. It is not warmed by *more sun*; it is warmed by the same sunlight handled differently. Replace T-shirts with oceans and snowfields and the stakes move from comfort to climate.

A historical anecdote makes the point with a different kind of immediacy. In the late nineteenth century, as industrial cities burned coal at a scale never seen before, the soot did not politely stay in the air above the factories. It settled downwind on buildings, on fields, and on snow. People noticed. Snow in and around cities could turn gray and dingy quickly; it melted sooner. Long before satellites, this was albedo written in grime. The physics was not yet formalized in the public imagination, but the experience was there: darkened snow does not behave like snow. In parts of the world where seasonal snowpack feeds rivers and reservoirs, earlier melt changes the timing of water supply, not because the total snowfall necessarily changes in the same way, but because the surface stops reflecting like it used to. The connection between color and water management is an odd but practical one, and it has been relearned many times.

Albedo also has a geographical personality. The most dramatic white-blue boundary on Earth is not a line you can draw on a political map but a moving frontier: the edge of sea ice and the seasonal reach of snow cover. At high latitudes in spring and summer, sunlight can pour in for long hours. Whether it is met by bright ice or dark water changes how much energy is stored locally. That energy does not always announce itself immediately as warmer air. It can go into melting ice, warming the top layer of the ocean, or shifting the timing of when the atmosphere feels the heat. But the first step is still the same: reflection versus absorption.

One temptation is to imagine albedo as something the surface controls entirely. In reality, the atmosphere stands between the sun and the surface like a semi-transparent curtain. Some sunlight is scattered or absorbed before it ever reaches the ground or ocean. Some of the light reflected by the surface is intercepted on its way back up. Clouds, especially, can dominate what space "sees" when it looks down. This matters for a subtle reason: a surface can undergo a dramatic change in reflectivity and yet the change in the *planetary* albedo can be muted because clouds and air molecules are already doing a large fraction of the reflecting. Think of painting a darker color on the floor of a room, then turning on thick fog machines: the floor still matters, but your view of it is filtered. This filtering does not make the surface irrelevant; it makes the accounting more complicated, and it helps explain why measuring Earth's reflectivity precisely has been a scientific project rather than a trivial one.

That project has its own quiet drama. Albedo is not measured by holding up a card and judging brightness. It is measured by carefully tracking how much sunlight arrives at the top of the atmosphere and how much leaves again after interacting with clouds, air, and surface. Satellites have made this global bookkeeping possible, turning the planet into something like a monitored engine whose inputs and outputs can be tallied. Even then, the answer comes with uncertainties, not because physicists have forgotten how to divide, but because the planet is restless. Clouds thicken and thin; aerosols spike after a volcanic eruption; sea ice retreats in summer and advances in winter; storms rearrange the patchwork. A global average must be coaxed out of a perpetually changing picture.

For intuition, it can help to imagine Earth as a household with a budget. The sun deposits a salary in the form of energy. The planet "spends" that energy by radiating infrared back to space. Albedo is the payroll department. It determines how large the salary is after taxes, before any spending decisions are made. Increase albedo and the take-home pay shrinks: less energy is absorbed, and the planet cools unless other terms change. Decrease albedo and the take-home pay grows: more energy is absorbed, and the planet warms unless compensated by increased infrared loss. This is not a metaphor meant to replace equations; it is a way to keep cause and effect straight when the details multiply.

The strongest albedo contrasts are not just between snow and ocean, but between *states* a surface can occupy. Water can be liquid and dark, or frozen and bright. Land can be forested and relatively dark, or covered in snow and bright. The same latitude can swing from reflective to absorptive as seasons turn. That makes albedo a fast lever, able to shift over days to months, and also a slow lever, able to shift over decades as the distribution of ice and vegetation changes. The word "mirror" is slightly misleading here, because mirrors are stable objects. Earth's reflectivity is a living mirror, changing with weather and with the slow churn of climate.

A concrete example brings the numbers to life. Imagine a square meter of surface under bright sunlight. Suppose about 300 watts of solar power reach that square meter after accounting for the angle of the sun and the filtering of the atmosphere (the actual number varies widely with location and time, but this is a reasonable, human-scale figure). If the surface is fresh snow with an albedo of 0.85, it reflects about 255 watts and absorbs about 45 watts. If the surface is open ocean with an albedo of 0.06, it reflects about 18 watts and absorbs about 282 watts. The difference in absorbed energy is roughly 237 watts per square meter—a change large enough to feel as heat if you could stand your hand on it. Spread across vast areas and long summer days, differences of this order are why the geography of whiteness matters.

The emotional punch of albedo, once you see it, is that it ties the fate of faraway places to ordinary experiences of light. The poles can sound like abstractions: latitudes, isotherms, contour lines. But a bright surface returning sunlight to space is a physical act with consequences. When that brightness is lost, the energy does not vanish; it stays in the Earth system, redistributed through air and ocean currents, stored and released, altering the background against which weather plays out. A child in a city may never set foot on sea ice, yet the color of the Arctic can still influence the climate envelope that shapes the child's world.

There is also a philosophical sting in how innocent the mechanism is. Albedo involves no exotic chemistry, no rare gases, no complicated biology. It is light hitting matter. Yet it is deeply entangled with everything else: with clouds that can either shield or trap, with the ocean that can hoard heat and deliver it later, with the land surface that can be brightened by snow or darkened by vegetation, and with human activity that can change surfaces directly by clearing forests or indirectly by warming the air that determines whether precipitation falls as rain or snow. The planet's mirror is not one object but a layered system, and each layer is adjustable.

Even the word "white" can mislead, because the whitest places are not always the most climatically powerful. A small patch of snow near the equator can be dazzling but brief; a broad seasonal snow cover across northern continents can be less brilliant per square meter but enormous in area. Similarly, bright clouds can reflect huge amounts of sunlight, but they are transient and variable. In terms of climate stability, some reflective features are reliable and some are fickle. The stability of the planetary mirror depends not just on how bright Earth can be, but on how persistent that brightness is.

One last detail is worth holding onto because it keeps the discussion honest. Albedo is only part of the energy story. A surface can reflect a lot of sunlight and still be involved in warming through other pathways. Snow, for instance, can insulate the ground beneath it, keeping soil warmer than bare, wind-scoured ice would. Clouds can cool by reflecting sunlight and warm by trapping infrared, with the balance depending on cloud type, thickness, and altitude. The ocean can absorb sunlight efficiently and then mix heat downward, delaying its effect on the atmosphere. These complications do not cancel the importance of albedo; they prevent it from becoming a cartoon villain or hero.

Still, as a first line of defense, albedo has a clarity that is rare in climate science. It translates the look of the world into a lever on temperature. The planet's mirror is not a metaphor bolted onto physics; it is physics visible to the naked eye. On a bright day, when sunlight flashes off a snowfield or dies into the dark of open water, the difference is the thermostat's quiet work: a choice, repeated billions of times each second across the globe, between sending energy back to space and keeping it close.

3.2 The Arctic Amplifier

The first time many people truly *see* Arctic change is not in a graph but in a photograph: a bear on a small floe, a ship pushing through waters that once held ice, a village perched above a coastline that looks as if it has been bitten away. The Arctic can feel like a distant stage set, but the physical engine of its rapid warming is startlingly simple. When a bright surface disappears, it stops sending sunlight back to space. The extra energy has to go somewhere, and in the Arctic it goes into water.

Sea ice is an odd material to build a climate system around. It is thin by geological standards, usually measured in meters rather than kilometers. It drifts, cracks, piles up, and melts from above and below. Yet it performs two enormous services at once. It is a lid between the ocean and the atmosphere, and it is a bright reflector in a part of the world that receives continuous sunlight for weeks to months each summer. Remove the lid and you do not merely reveal dark water; you also rearrange the schedule of when and where heat is stored.

The simplest version of the story is the *ice—albedo feedback.* It is a positive feedback, meaning it reinforces an initial change rather than resisting it. Warming melts ice. The newly exposed ocean is darker than ice and absorbs more sunlight. The additional absorbed energy warms the ocean further, which melts more ice. The loop can begin with slightly warmer air, slightly warmer water, a few more clear-sky days, or a shift in winds that pushes ice away from a region. Once the ocean is exposed, sunlight does much of the work.

The words "positive feedback" sometimes sound like praise, as if the climate were being encouraged. In this context, "positive" means arithmetic: the sign is positive. The feedback adds to the initial warming. It is the opposite of a stabilizing mechanism. And it is fast. The moment a patch of ocean is exposed to summer sun, its energy budget changes immediately.

A person can grasp the strength of the effect with an image from daily life. Picture a black asphalt parking lot after a snowstorm. While the snow remains, the lot stays bright and cool, even on a sunny day; the sunlight bounces off. When cars and plows scrape the snow away, the same sunlight suddenly becomes heat you can feel radiating from the ground. The Arctic is like that, except the "parking lot" is a moving ocean, and the plowing is done by temperature, waves, and wind.

One reason the Arctic is especially sensitive is geometric. The pole is surrounded by continents, but the center is ocean. When ice retreats, it reveals a vast, contiguous dark surface. Another reason is seasonal timing. Arctic summer brings near-continuous daylight, so any change in reflectivity during that period can accumulate energy for weeks. The Arctic winter is dark, but the energy saved or spent in summer does not vanish when the sun sets; it is banked in the ocean and paid out later.

That seasonal banking is where the story becomes more interesting than a simple mirror. Many people imagine the strongest warming should occur in summer, when sunlight is abundant and ice melt is visible. Yet observations and models often show that the most dramatic Arctic surface warming happens in autumn and winter. This is not a contradiction; it is the signature of a system with memory.

When sea ice is present, it insulates the relatively warm ocean from the cold air above. It slows the ocean's loss of heat in winter, and it maintains a stable boundary that encourages temperature inversions: cold air near the surface, warmer air above. When sea ice is lost in summer, the ocean absorbs extra sunlight and warms. As autumn arrives and the air cools, that open water releases its stored heat upward. Heat fluxes from the ocean into the atmosphere can surge, warming the lower atmosphere and delaying the refreezing of ice. The energy that entered through a change in summer reflectivity emerges as winter warmth. The feedback has a seasonal rhythm: sunlight enters when the sun is high; warmth lingers and is expressed when the atmosphere is most sensitive.

A concrete example makes this easier to picture. Imagine a stretch of ocean that used to be ice-covered through most summers. In the older state, sunlight might have struck bright ice and snow and been largely reflected. In a newer state, the same sunlight strikes open water for a larger part of the season. The ocean takes in extra energy not because the sun has changed, but because the surface has. That additional energy is mixed into the upper ocean by wind and waves. When the sun dips lower and the air cools, the ocean does not instantly give up its heat. It releases it over weeks to months, partly as sensible heat (warming air directly) and partly as latent heat through evaporation (water vapor carries energy upward and later releases it when it condenses). The surface air temperature, measured where people live and where ecosystems function, responds strongly to these autumn and winter heat releases. The Arctic warms in the dark because the ocean was charged in the light.

The most famous phrase for the Arctic warming faster than the rest of the world is *Arctic amplification*. The term is not a claim that the Arctic has its own special greenhouse gases; it is a description of a pattern: Arctic surface temperatures rise more rapidly than the global average. Albedo feedback is a major part of that amplification, but the full picture is an interlocking set of feedbacks and adjustments.

One of the more counterintuitive ones is the *lapse-rate feedback*. "Lapse rate" is the rate at which air temperature decreases with height. In many places, warming tends to be stronger aloft than at the surface, partly because the tropics are shaped by deep convection that mixes heat upward. In the Arctic, warming can be strongest near the surface, especially in winter, because stable layers and temperature inversions trap heat close to the ground. When the surface warms more than the air above, the atmosphere radiates infrared to space less efficiently than it otherwise would, which acts as a positive feedback. This is technical, but the intuition is simple: in a strongly stratified atmosphere, where layers do not mix much, warming near the bottom has an outsized effect.

There is also the *Planck response*, named after the physics of thermal radiation. Any warm object emits more infrared radiation as it warms. This is the climate system's most basic stabilizing tendency: warm the planet, and it naturally radiates more heat away. But in the Arctic, because warming is concentrated near the surface and because of the structure of the atmosphere, the effective strength of this stabilizing response can differ from lower latitudes. The result is not a different law of physics but a different expression of the same law in a particular atmospheric setting.

Clouds are the wild card. Over bright ice and snow, clouds can sometimes warm more than they cool, because their ability to trap infrared can outweigh the sunlight they reflect—especially when the surface below is already reflective. Over dark open water in summer, clouds can cool strongly by reflecting sunlight that would otherwise be absorbed. In winter, when there is little sunlight to reflect, clouds tend to warm by acting as an infrared blanket. Changes in cloud cover and cloud type, influenced by open water and by shifts in weather patterns, can either enhance or partly offset the albedo-driven changes. The uncertainty is not merely scientific; it is physical. Clouds are both mirrors and blankets, and the balance shifts with season and circumstance.

For all that complexity, a strong, modern result from climate model analysis is conceptually clean: large sea-ice loss tends to reorganize the whole Arctic feedback environment. In experiments where sea-ice decline is suppressed or reduced, several other feedbacks weaken markedly, suggesting that sea ice is not just one player among many but a central switch that changes the operating conditions for the rest of the system. Water vapor feedback, by contrast, changes relatively little, because the Arctic's cold air holds less moisture and because much of the water vapor story is global. The Arctic's special vulnerability is therefore not primarily a tale of humidity. It is a tale of surfaces and seasons.

A historical moment that brought the Arctic amplifier into public view was the late-summer sea-ice minimum of 2007. Scientists had been tracking sea ice from satellites since 1979, and the downward trend was clear, but 2007 produced a dramatic plunge that startled even researchers accustomed to bad news. Weather played a role: winds and patterns of pressure helped compact and export ice, and clear skies favored melt. But the event was also a lesson in preconditioning. Thin ice, formed over previous winters, is easier to destroy in summer. Open water created early in the melt season absorbs more sunlight, which creates more open water. A single summer became a kind of demonstration experiment in how quickly the ice-ocean system can shift state. The years after 2007 did not simply return to an earlier "normal." Variability continued, but the baseline had moved.

The 2012 minimum deepened that lesson. Again, weather contributed, including a powerful storm that stirred the ice. Yet storms do not create melt energy out of thin air. They redistribute ice, break it up, and enhance mixing; they can speed the conversion of absorbed heat into lost ice. The fundamental vulnerability was that there was less thick, multi-year ice to begin with. Older Arctic ice is more resilient, not because it is magically colder, but because thickness is stored winter cold. Losing multi-year ice is like losing the savings account that helps you endure a bad month.

The human impacts of these changes can feel indirect from the perspective of a mid-latitude city, but in Arctic communities they are visceral. Sea ice is not merely scenery; it can be a travel corridor, a hunting platform, and a protective buffer against waves. When ice forms later and breaks up earlier, the window for safe travel shrinks. When storms hit open water where ice once damped waves, coastal erosion can accelerate. Infrastructure built on assumptions about frozen seasons and stable shorelines becomes harder to maintain. The feedback loop that begins as a change in reflectivity ends up as a change in daily logistics.

There is an ecological dimension as well. The timing of sea-ice retreat influences the timing and location of phytoplankton blooms, because sunlight can penetrate open water and stimulate growth. Fish distributions can shift. Marine mammals that rely on ice for resting, breeding, or hunting face altered habitats. None of these outcomes is controlled by albedo alone, but albedo helps set the stage by determining where the ocean becomes available to light and heat.

A subtle but crucial point is that not all "ice" is equal in the albedo story. The surface condition matters. A slab of sea ice covered in fresh snow can be intensely reflective. The same slab, after weeks of summer melt, can have a surface dotted with ponds that are nearly as dark as the ocean. From above, the ice becomes a mosaic of white and blue long before it breaks apart. That means the albedo feedback is not solely about how much area is ice-covered; it is also about the texture, wetness, and cleanliness of the ice surface. Small changes in melt onset, snowfall on ice, or the persistence of melt ponds can alter how much sunlight is absorbed even if the ice extent, measured in square kilometers, looks similar.

That distinction matters because it discourages a comforting simplification: "as long as some ice remains, the mirror remains." The mirror can become smudged. The Arctic can be ice-covered but darker than it used to be.

Another temptation is to treat the Arctic as a self-contained box. In reality, it is connected to the rest of the planet by winds and currents that move heat poleward. When sea ice retreats and the Arctic warms, these transports can adjust. Some of the warming comes from local absorption of sunlight, and some comes from altered flows of heat and moisture from lower latitudes. There is debate about the degree to which Arctic changes influence mid-latitude weather patterns, in part because the atmosphere is noisy and because many drivers operate simultaneously. What is clearer is that the Arctic is not merely receiving change; it is participating in a global redistribution of energy.

The speed of the sea-ice response is part of why it has become such a prominent symbol. A forest can take decades to migrate; an ice sheet can take centuries to reorganize. Sea ice can retreat dramatically in a single season. That does not mean the system is finished changing after a single summer. It means that a fast component has been loosened, and fast components have a way of dragging our attention to the present tense.

Sea ice is also a kind of psychological amplifier. It is visible. It is binary in a way that gradual warming is not: there is ice, or there is water. The boundary is sharp. In satellite images, the change reads like a disappearing continent. People instinctively understand that a loss of something white implies a gain of something dark, even if they have never heard the word albedo. In that sense, the Arctic teaches planetary physics by providing a high-contrast diagram of it.

At the end of a summer, when sunlight has already done its work, the ocean carries extra heat into the darker months. The first thin ice that tries to form is fragile. It grows by freezing at the bottom, while snow on top can insulate it from the frigid air. Leads—cracks of open water—release heat like vents. Each patch of open water is small, but together they make an archipelago of warmth that the atmosphere cannot ignore. The old Arctic, with thicker ice and fewer openings, behaved more like a sealed cooler. The newer Arctic behaves more like a pot left slightly uncovered on a stove: heat escapes easily, steam rises, and what happens above the surface is shaped by what is happening below.

That is why the Arctic can warm quickly even if the global forcing changes only gradually. A small, steady push on temperature can cross a threshold in the seasonal cycle: earlier melt, later freeze, thinner ice, more open water at the time of maximum sunlight. Once the calendar shifts, the physics follows the calendar. The feedback does not require drama. It requires timing.

There is a final image worth keeping, because it captures both the elegance and the danger of the Arctic amplifier. Imagine the Arctic Ocean in midsummer as a vast solar collector. The collector is designed, by nature, to have a reflective cover: sea ice and snow. When the cover is intact, much of the incoming energy is bounced away and the collector stays relatively cool. When the cover is punctured, dark water is exposed, and the collector does what collectors do: it gathers energy efficiently. The collected energy does not sit politely in place; it is distributed through currents and released to the air when conditions allow. A mirror has become an absorber, and the Arctic responds not with a gentle adjustment but with an amplified answer, spoken most loudly when the sun is gone and the warmth seems, at first glance, to come from nowhere.

3.3 The Slumbering Giants

Sea ice can vanish with shocking speed, as if the ocean has simply decided to stop wearing its white hat. Continental ice sheets do not behave that way. Greenland and Antarctica are not seasonal skins but immense bodies of frozen water, thick enough to bury mountains and heavy enough to press the crust beneath them down. They respond slowly for the same reason a cathedral cools slowly after the doors are opened on a winter night: there is so much material, and so much stored cold, that a brief change in weather is an insult rather than an injury. Yet the same mass that makes them sluggish also makes them consequential. When these giants shift, the planet's coastline shifts with them.

The key idea is *thermal inertia*—the tendency of a system to resist rapid temperature change because it can store a large amount of heat (or cold). Oceans have high thermal inertia, which is why coastal climates are milder than continental interiors. Ice sheets also have high thermal inertia, but their inertia is not simply a matter of temperature. It includes geometry. An ice sheet is a moving pile of solid water that flows, very slowly, under its own weight. The surface climate can warm quickly, but the interior of a thick ice sheet can remain far below freezing for centuries. The surface can melt or lose snow in a warm summer, while the deep ice retains a memory of colder centuries, layered like pages in a book.

That stored cold is one reason it is easy to underestimate ice sheets by looking only at day-to-day experience. An unusually warm week does not melt Greenland. Even an unusually warm decade does not melt Antarctica. The most important changes often begin at the margins, where ice meets ocean and rock, and where the physics is less like a frozen parking lot and more like a slow battle between a glacier and a moving sea.

It helps to start with scale. Greenland's ice sheet is roughly a continental dome, thickest in the center and thinner at the edges, feeding glaciers that spill into fjords. Antarctica is a different animal: much of its ice rests on bedrock below sea level, held back by floating ice shelves that extend into the ocean like broad tongues. Both systems contain enough water to raise global sea level by many meters if lost, but they do not lose mass by simply warming up like ice cubes in a glass. They lose mass through a combination of surface melting, snowfall changes, iceberg calving, and melting where ocean water attacks the ice from below. A slow, enormous system can still have fast, fragile points, and those points often sit where ice touches seawater.

A historical anecdote captures the power of ice-sheet memory. During the nineteenth century, engineers and surveyors began to measure the slow rebound of land in places that had been covered by ice during the last glacial period. Scandinavia and parts of Canada were rising, millimeter by millimeter, as if the Earth itself were taking a deep breath. The explanation, developed over time through geophysics, is that the crust and mantle deform under the weight of ice and then relax upward when the load is removed. This is not climate change in the everyday sense; it is the planet still responding to ice that vanished thousands of years ago. The land's slow rise is a reminder that ice sheets do not merely respond to climate. They reshape the solid Earth, and the solid Earth, in turn, reshapes how ice flows.

That interplay is part of why ice sheets can be both stable-looking and vulnerable. For long periods they can sit near a balance: snowfall accumulates on top, ice flows outward and is lost at the edges, and the overall size changes slowly. But balance does not mean safety. A system can be balanced on a plateau or balanced near a cliff. The challenge is that the cliff edge is not always visible from above.

The simplest stabilizing picture is a household one. Snowfall is income; melting and ice discharge are expenses. If expenses rise a little, the household draws down savings. If expenses rise for long enough, the savings run out. Ice sheets do have a kind of savings: their thickness and their internal cold. But they also have an ongoing set of expenses that can increase in ways that are not gradual. Warm water can reach places that were once protected. A floating ice shelf can thin, weakening its ability to restrain the glacier behind it. A grounding line—the point where ice stops resting on bedrock and begins to float—can retreat inland. If the bed beneath slopes downward inland, retreat can expose thicker ice to seawater, increasing flow and encouraging further retreat. The geometry can generate its own momentum.

This is one of the most unsettling aspects of ice-sheet physics: the most dangerous feedbacks are not necessarily about temperature directly, but about shape and contact. A modest change in ocean conditions can trigger a rearrangement in how ice meets water, and that rearrangement can alter the rate at which ice is lost.

Consider the Antarctic ice shelves. They are already floating, so when they melt they do not directly raise sea level, just as melting sea ice does not raise sea level. Their importance is mechanical rather than volumetric. An ice shelf acts like a brace, providing *buttressing* that slows the glaciers feeding it. Thin the shelf and you weaken the brace. Remove the shelf and the glacier behind it can accelerate, sending more grounded ice into the ocean, which *does* raise sea level. The shelf is a gate, and its strength controls the flow through the gate.

A vivid modern example is the Larsen B Ice Shelf on the Antarctic Peninsula. In 2002, a large portion of it disintegrated over a matter of weeks, breaking into a swarm of icebergs. The dramatic collapse was preceded by surface melting and the formation of melt ponds. Water filling cracks can pry ice apart—a process called hydrofracture—and once the shelf is fragmented, it loses its ability to behave as a coherent brace. After the shelf's collapse, the glaciers that had fed it sped up and thinned. The shelf itself did not add to sea level when it shattered, but the response of the land-based ice behind it did. The event was a lesson in how a slow system can still contain brittle failures.

Greenland offers a different kind of drama. There, surface melt plays a larger role. Meltwater can pour into crevasses and reach the base of glaciers through moulins, lubricating the interface between ice and bedrock and briefly speeding up flow. Some of these effects are episodic, tied to melt seasons, and the long-term implications are still an active area of study. But Greenland's most sobering contribution is simpler: when air temperatures rise, the area and duration of surface melt can expand. Darkening of the ice surface by dust, soot, and biological growth can lower albedo, allowing more sunlight to be absorbed during summer. The surface of a gigantic ice sheet, once thought of as uniformly white, can become patchy and more absorbent. Even there, though, the response is not a clean, linear melt. Some meltwater refreezes within the snow and firn (the compressed, granular layer that sits between fresh snow and solid ice), temporarily storing water without contributing to sea-level rise. Over time, as refreezing capacity is overwhelmed or as firn becomes denser, more meltwater runs off. The ice sheet has internal plumbing, and the plumbing changes as the climate changes.

Thermal inertia shows up in these stories in several ways. The cold interior delays deep warming, meaning that some processes are initially limited by how quickly heat can penetrate. But the same inertia can also create a misleading calm: the surface can look stable while boundaries are being altered at the edges. It is possible for an ice sheet to be committed to long-term change even if the present-day rate looks manageable, because the system has been nudged onto a trajectory that plays out over centuries.

This is where language becomes tricky. People naturally ask whether ice sheets have *tipping points*. A tipping point suggests a threshold beyond which change becomes self-sustaining and difficult to reverse. The idea is not science fiction. Certain feedbacks in ice sheets can, in principle, create threshold behavior. But there is also a danger in treating tipping points as a single dramatic moment, like a dam breaking. Some thresholds are more like passing the last exit on a highway: the car continues smoothly, but the route back becomes long and costly.

Recent work has highlighted a mechanism that makes this threshold idea more concrete for the ocean-facing margins. Warm ocean water can intrude into the grounding zone, the delicate region where ice transitions from grounded to floating. If that intrusion increases melting, it can widen or reshape the cavity that allows water to enter, which can permit even more warm water intrusion and further melt. In idealized modeling, small increases in ocean temperature can push the system into a regime where intrusion and melting reinforce one another strongly, producing a kind of runaway retreat in the grounding zone geometry. The unsettling implication is not that all of Antarctica is about to collapse, but that some parts may be more sensitive to ocean warming than models that smooth over grounding-zone details might suggest.

At the same time, there is genuine debate about how close current ice-sheet geometries are to such thresholds, and about which mechanisms dominate in specific regions. Some modeling work argues that present-day grounding lines may not be undergoing certain classic instabilities under steady conditions, even while acknowledging that future forcing could push them into a less stable state. This is not scientific hedging for its own sake; it reflects the fact that the Antarctic margin is not one uniform machine. It is a collection of basins with different bed shapes, different ice-shelf geometries, different exposure to warm deep water, and different histories. A blanket statement about "Antarctic instability" is therefore less informative than a map.

The slow timescale of ice sheets also creates a moral hazard for human attention. Sea-ice loss can be watched year by year. Ice-sheet mass loss accumulates with the steadiness of compound interest, and the consequences show up most clearly in a number most people do not think about until it reaches their doorstep: sea level. A few millimeters per year sounds small until you multiply it by decades, add storm surges, and recognize that the baseline from which waves attack a coastline has risen. The most vulnerable places are not necessarily those with dramatic cliffs being gnawed away, but low-lying deltas and islands where the difference between nuisance flooding and unlivable flooding is measured in centimeters.

There is another slumbering giant often discussed alongside ice sheets, and it belongs to the same high-latitude world even though it is not white. Permafrost—soil that stays frozen for at least two years, often much longer—stores vast quantities of organic carbon. In cold regions, plant material can accumulate faster than microbes can decompose it, and freezing locks it away. When permafrost thaws, microbes wake up and begin converting that stored carbon into carbon dioxide and methane. This is an amplifying feedback: warming leads to thaw, thaw leads to greenhouse gas release, greenhouse gas release leads to more warming.

Permafrost illustrates a different kind of inertia. Like ice sheets, it has memory. Heat must diffuse into the ground; snow cover can insulate soil; vegetation can shade it; water can carry heat. Some thaw is gradual, deepening the "active layer" of soil that melts each summer. But permafrost can also fail abruptly. When ice-rich ground thaws, it can collapse into hollows, forming thermokarst landscapes of slumps, ponds, and destabilized hillsides. Whether thawed ground becomes wet or dry affects which gases dominate: oxygen-rich, drier soils tend to favor carbon dioxide; waterlogged, oxygen-poor conditions can favor methane. Wildfires can remove insulating vegetation and darken surfaces, speeding thaw. This is not a single lever but a set of interacting ones, and mechanistic models show that local hydrology and topography can control outcomes. Two nearby slopes can respond very differently to the same air temperature change.

Permafrost carbon is not ice-sheet sea-level rise, but the pairing is instructive. Both are slow systems with large potential. Both contain nonlinear pathways. Both respond to warming in ways that are heterogeneous across space and time. And both can turn a regional change into a global effect: one through sea level, the other through atmospheric composition.

Returning to ice sheets, it is tempting to ask a binary question: stable or unstable? The more accurate question is how stability changes with forcing, and how quickly. Ice sheets can be stable under one climate and unstable under another, and the transition between those regimes may not be obvious until the system has moved partway through it. The old view of ice sheets as inert piles has been replaced by a view of them as dynamic systems with moving boundaries and internal reorganizations. Yet the "slumbering" metaphor still holds in a crucial sense: their biggest consequences unfold on timescales longer than a news cycle and longer than an election cycle.

There is, however, a kind of immediacy hidden inside that slowness. Because ice sheets respond slowly, decisions made over decades can commit the planet to changes that play out over centuries. The children of the twenty-first century may live with the early stages of those commitments; their descendants may live with the coastlines those commitments produce. This is not a prediction of one inevitable outcome but a description of how long-lived systems work. Once a large body of ice is set on a path of retreat, cooling the air back down does not instantly rebuild it, just as turning down the thermostat does not instantly refreeze a lake that has already warmed through summer. The system needs time, and time is not evenly distributed among generations.

One way to picture the ice sheets' inertia is to imagine them as slow clocks built into the planet. Their ticking is measured in the creep of ice, the migration of grounding lines, the thinning of shelves, the slow conversion of snowfall into ice and ice into water. The hands move so gradually that a single glance can fool you into thinking nothing is happening. But a sequence of glances, spaced over years, reveals motion that cannot be unseen.

In the deep cold of Antarctica's interior, snow still falls and compacts, layer upon layer, preserving traces of ancient atmospheres. At the margins, where the ice meets the sea, the planet negotiates with saltwater that is not obliged to respect the old boundaries. The negotiation is conducted in melt rates, fracture lines, and shifting points of contact. It is quiet work, but it is decisive, and it leaves its signature not only in the polar landscape but in the height of the ocean that laps at every shore.

Chapter 4

The Memory of Water

If the atmosphere is Earth's mood, the ocean is its temperament; vast, slow to change, and governing the long-term stability of the system. We descend beneath the waves to investigate how the ocean's immense thermal mass masks the immediate severity of warming while simultaneously committing the planet to centuries of climate adjustment.

4.1 The Hidden Reservoir of Heat

If Earth had a single, easily readable thermometer, climate change would be simpler to argue about. The mercury would rise, we would nod, and the story would end. Instead, the planet keeps its main ledger somewhere less theatrical than the air: in the ocean, written as heat stored layer by layer. The atmosphere is where we *feel* weather and measure headlines, but it is not where most of the extra energy goes.

That imbalance between feeling and accounting begins with a mundane property of matter: specific heat capacity, the amount of energy needed to raise the temperature of a kilogram of a substance by one degree. Water's specific heat is high. Put plainly, water is stubborn. It takes a great deal of energy to warm it, and once warm it takes a long time to cool. This is why a coastal city can have a mild evening while an inland city at the same latitude swings from hot to cold in a single day. Land heats and cools quickly; the sea changes its mind slowly.

Scale that up from a beach to the planet and the ocean becomes something like a thermal flywheel. Greenhouse gases tilt Earth's energy balance so that more energy arrives than leaves. The first place much of that surplus ends up is the ocean surface, because the ocean covers about seventy percent of the planet and is in direct contact with the air. Wind and waves stir the surface, and turbulence pushes heat downward. Over time, currents and mixing distribute it into deeper layers. What looks like a modest change in sea surface temperature can correspond to an enormous gain in total ocean heat, because temperature is only the visible tip of a much larger volume.

A useful mental image is a pot of soup on a stove. If you only taste the surface, you might think it is barely warming, especially if you keep stirring in cool ingredients. But the stove is still on, the total energy is still increasing, and before long the whole pot is warmer. The ocean is Earth's soup pot: deep, well-mixed in places, stratified in others, and vast enough that the energy we add each day can hide in it for decades without giving the surface an obvious jolt. When the surface *does* jolt—during an El Niño year, for example—the jump feels dramatic precisely because our senses are tuned to air and skin, not to the slow arithmetic of a global heat reservoir.

Scientists have tried to express this disparity in a way that does not depend on specialist intuition. One popular comparison is to translate the planet's ongoing energy gain into the equivalent of detonating several Hiroshima-sized bombs per second, continuously, in the climate system. The point is not the exact number (which depends on the time period and estimate), but the mismatch between the violence of the energy flow and the relatively small wiggles in global mean surface temperature from year to year. The ocean can absorb a huge amount of energy with only a small rise in temperature because of its mass and heat capacity. If the atmosphere carried the same fraction of the added heat, the air temperature changes would be far larger, far faster, and far harder to ignore.

The ocean's dominance in the heat budget is not a poetic claim; it is an engineering fact. Consider the thickness of the relevant layers. The lower atmosphere that matters most for surface climate is measured in a few kilometers. The actively exchanging upper ocean is hundreds of meters thick, and below that lies a deep ocean whose average depth is nearly four kilometers. Even if only the top two kilometers are counted—a common benchmark for observation products because it is where measurements are best and where much of the recent heat has gone—that is still a column of water vastly heavier than the column of air above it. Mass matters. A small temperature change multiplied by an enormous mass becomes a formidable energy change.

This is where the lag comes from. When greenhouse gases increase, they do not heat the air alone; they alter the rate at which Earth can radiate heat back to space. The planet responds by warming until outgoing radiation again matches incoming sunlight. But reaching that new balance is like warming a house with thick stone walls. Turn on the heater and the air warms quickly, giving the impression that the job is mostly done. Hours later the walls are still absorbing heat, and the room continues to adjust. Turn off the heater and the air cools quickly, but the walls release their stored warmth and the room does not drop to its original temperature right away. The ocean is the stone wall of the climate system. It does not simply track the atmosphere; it *governs* the pace at which the whole system can settle.

There is a second reason surface temperatures can mislead: the ocean does not store heat evenly. The top few meters respond quickly to sun, wind, and rain. The upper few hundred meters are stirred by seasonal mixing and storms. Below that, the ocean becomes more layered, with density stratification (set by temperature and salinity) resisting vertical mixing. Heat can be pushed down by particular processes—subduction along currents, eddies that pinch off and sink, or deep winter mixing in stormy regions—but the average pace is slow. This creates a world where the surface can have a hot year or a cool year while the deeper reservoir continues its steady accumulation, like a bank account that grows even when daily spending fluctuates.

A historical anecdote helps here because it shows how recently this hidden reservoir became visible. For much of the twentieth century, ocean temperature measurements were opportunistic: sailors took bucket samples, ships lowered thermometers, and later expendable bathythermographs (XBTs) were tossed overboard to record a temperature profile as they fell. These data were invaluable, but patchy. Large regions and seasons were poorly covered, and different instruments had different biases. The ocean's heat gain was real, but the observing system was not built to weigh it.

That changed around the turn of the millennium with a project that deserves to be as famous as a moon landing, even though it happened quietly in rough seas. The Argo program began deploying autonomous profiling floats across the global ocean. Each float drifts at depth, then every ten days or so it sinks and rises, measuring temperature and salinity as it ascends and sending data back via satellite. Today, thousands of Argo floats create a living, roaming network that turns the ocean from a mystery into a monitored organ of the climate system. The first time a global Argo map appeared, it was not a single dramatic curve but a new kind of confidence: the sense that the planet's main heat storehouse could finally be audited, not guessed.

Auditing matters because the ocean is also where delayed consequences build up. Heat stored in water expands it, raising sea level even without adding new water from melting ice. Heat stored in the upper ocean can later be released to the atmosphere during particular circulation patterns, shaping regional climates. And heat pushed into deeper layers is not gone; it is simply moved onto a slower clock. The phrase "committed warming" is sometimes used for this, but it can sound abstract. A more tangible phrasing is that the ocean acts like a thermal battery. As long as the planetary energy imbalance remains positive, the battery charges. Even if the imbalance were reduced, the battery would not instantly discharge; it would continue to influence the surface because the ocean and atmosphere are coupled, exchanging energy and moisture.

The scale of this hidden heat is best shown visually. The exact numbers depend on the dataset and the period chosen, but the qualitative picture is robust: the ocean, especially its upper layers, holds the overwhelming majority of the added energy since the late twentieth century, while the atmosphere's share is a thin sliver. The chart below illustrates that contrast using representative cumulative values since 1970, separating the ocean into two layers to emphasize depth.

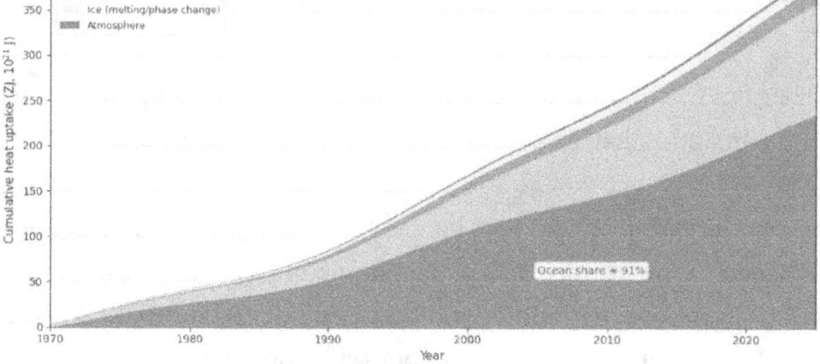

One way to read that plot is to ask a deceptively simple question:

if the atmosphere holds only a tiny fraction of the accumulated heat, why does the atmosphere get so much attention? Because it is the part of the system we inhabit. Air temperature changes govern heat waves, crop stress, wildfire risk, and the limits of human comfort. But air temperature is also noisy, easily nudged by short-term variability. The ocean, by contrast, is where the climate system keeps the long-term trend with fewer distractions. Measuring ocean heat content is like weighing a ship rather than watching the waves around it.

This helps resolve a common confusion: the apparent mismatch between "pause" narratives and continued warming. There have been periods, lasting several years, when the rise in global mean surface temperature slowed relative to the years immediately before. Some of those slowdowns were connected to volcanic aerosols, some to changes in solar output, and many to internal variability such as La Niña conditions that encourage the ocean to take up more heat in certain regions and layers. During such intervals, the planet's overall energy gain did not vanish; it was redistributed, with a larger fraction going into the ocean while the surface warmed more slowly. When the variability flips, some of that stored ocean heat can contribute to a burst of surface warmth. The ocean's thermal inertia does not make the surface irrelevant; it makes the surface a lively messenger of a deeper process.

Thermal inertia also changes how we think about "stopping" warming. Imagine that greenhouse gas concentrations stabilized tomorrow. The planet would still be out of equilibrium for a time, because the ocean would still be catching up to the altered radiation balance. The surface might not immediately stop changing; the deep ocean would continue to adjust, and the redistribution of heat would continue. This is not a counsel of despair. It is a statement about timescales. Human actions can change the direction of forcing quickly; the ocean changes the pace at which consequences unfold.

That is why policies aimed at limiting peak warming focus so intensely on cumulative emissions and near-term reductions: they are attempts to stop adding to the battery before it charges further.

There is a more intimate way to see the ocean's memory. Stand on a pier on a cool spring day and watch the sea smoke slightly as warmer water meets colder air, or feel how a late-summer ocean keeps the evening warm even after the sun drops. Those sensations are local expressions of the same physics that operates globally. Water stores heat and shares it slowly. It smooths extremes, but it also preserves a record. The ocean you touch today contains, in diluted form, the energy imbalance of yesterday's atmosphere.

The hidden reservoir of heat is not hidden because it is secret; it is hidden because it is inconveniently large and slow, resisting the tempo of daily attention. A politician can point to a cold snap; an ocean cannot be waved away so easily. It responds to the energy budget with steady patience, taking what the atmosphere cannot hold, and in doing so it keeps the future tethered to the past.

4.2 The Thousand-Year Loop

A cup of tea cools because warm water rises and cool water sinks, setting up a small circulation that keeps the drink from stratifying into a hot layer on top and a cold layer below. The ocean does something similar, but with a twist that makes it one of the great timekeepers of the climate system: it is not temperature alone that controls whether seawater floats or sinks. Salt matters too.

Thermohaline circulation is the name given to large-scale ocean motion driven by *thermo* (heat) and *haline* (salt) effects on density. Density is simply mass per unit volume, but it is the quiet dictator of the ocean's vertical structure. Warm water is usually less dense than cold water, so it tends to sit on top. Saltier water is denser than fresher water, so it tends to sink. Put those together and you get a delicate contest: polar regions cool surface waters, encouraging sinking; rainfall and ice melt freshen surface waters, discouraging sinking. The outcome of that contest helps determine where the ocean opens a trapdoor to the deep.

The most famous part of this story is the Atlantic Meridional Overturning Circulation, often shortened to AMOC. "Meridional" means north—south, and "overturning" captures the essential idea: warm water flows northward near the surface, releases heat to the atmosphere, and in a few key places becomes dense enough to sink, returning southward at depth. When this loop is strong, it moves a remarkable amount of heat out of the tropics and into the North Atlantic. That heat transport helps explain why western Europe can have relatively mild winters for its latitude compared with, say, eastern Canada. It also helps explain why disruptions to the loop are taken so seriously: this is not a decorative current painted on a textbook map, but a major conveyor of energy, freshwater, and carbon.

It is easy to picture this circulation as a giant river in the sea. That picture is comforting because rivers have banks and clear directions. The real ocean is more like a crowded city with many streets: currents branch, merge, swirl into eddies, and detour around underwater mountains. Yet the broad outline of the overturning loop is robust enough to be described without lying. In the Atlantic, warm, salty surface waters move northward. As they travel, they lose heat to the air, especially in winter storms. Cooling increases their density. In the Labrador and Nordic Seas, where winds and cold air can strip heat rapidly, surface waters can become dense enough to sink. Once they sink, they join deep currents that flow southward, eventually connecting into the global deep ocean. Far away, in other regions—notably the Southern Ocean—deep waters are brought back toward the surface through a combination of wind-driven upwelling and mixing. The loop closes when those resurfaced waters are warmed and rejoin surface currents.

Calling it a "thousand-year loop" is shorthand for the timescale involved. A parcel of water that sinks into the deep North Atlantic does not pop back up next week. Much of it will spend centuries in the interior ocean. Some pathways are faster, some slower, and the ocean is not made of labeled parcels moving politely along a conveyor belt. But the basic point survives all the complexities: deep circulation is slow enough that it gives the climate system memory measured in centuries and, in places, closer to a millennium.

That memory is not mystical. It is mechanical. Once heat is transferred from the surface into deeper layers, it is insulated from direct contact with the atmosphere. The upper ocean can exchange energy with the air in days to months. The deep ocean is separated from the air by density stratification and the sheer distance from the surface. To change the deep ocean quickly, you would need either rapid vertical mixing everywhere (which does not happen) or a dramatic change in the sites where deep waters form and sink. Without that, the deep ocean changes on its own schedule, integrating many years of surface conditions into a slower, smoother response.

The slowness also creates a conceptual trap. People sometimes imagine that if the ocean mixes slowly, it must be safe from human influence. In fact, slowness is precisely what makes the ocean's influence so consequential. A fast system reacts quickly and can also recover quickly. A slow system spreads the consequences across generations. Heat that slips beneath the surface is not immediately harmful in the way a heat wave is harmful, but it is not harmless. It expands seawater, raising sea level. It shifts the balance of ice shelves from below. It changes the conditions in which marine life evolved. And it can return, gradually, to the surface, nudging surface temperatures and weather patterns long after the original atmospheric forcing has changed.

The ocean's long loop can be inferred in several ways, but one of the most vivid is to give seawater an "age." This is not the age of the molecules (water molecules are constantly recycled) but the time since that water last touched the atmosphere. A surface parcel in contact with air has access to oxygen, carbon dioxide, and a suite of chemical tracers. When it sinks, it is cut off from that contact and begins to drift chemically. Oxygen gets consumed by respiration; certain isotopes decay; carbon accumulates in different forms. By measuring these tracers, oceanographers can estimate how long it has been since the water was ventilated, meaning last exposed to the atmosphere. The numbers are arresting: in much of the deep ocean, ventilation ages are hundreds of years. In the deep Pacific, some waters can be closer to a millennium in ventilation age, and in some reconstructions and simulations the oldest waters can be older still.

This is where science becomes both detective story and cautionary tale. Radiocarbon, the famous isotope used to date archaeological finds, is also used in the ocean. It enters seawater through exchange with the atmosphere. Once water leaves the surface, radiocarbon decays with a known half-life. Measure how much remains and you can infer how long the water has been away from air. It sounds straightforward until you remember that the ocean does not begin with a perfectly uniform radiocarbon "timestamp." Surface waters have different starting values depending on where they formed, and the atmosphere itself changes over time. Lingwei Li and Zhengyu Liu (2024) emphasized that radiocarbon-based ventilation ages can be biased relative to the true transit time because the starting signature changes and because reservoir ages—the difference between atmospheric radiocarbon and ocean surface radiocarbon—vary in time and space. In other words, the ocean does not come with a single stopwatch. It carries a set of clocks whose faces can be smudged by the very processes one is trying to measure. The existence of bias does not erase the thousand-year memory; it reminds us that pinning down exact numbers is hard in a system that is both moving and chemically active.

A concrete example makes the stakes clearer. Imagine a patch of surface water in the Nordic Seas in winter. Cold winds roar off the ice edge. The sea surface loses heat quickly. If that patch is also salty enough, it becomes dense and sinks. As it sinks, it takes with it a snapshot of surface conditions: the temperature of that year, the dissolved gases, traces of pollutants, even the isotopic composition shaped by rainfall patterns and evaporation. Once in the deep, it joins a slow river of water sliding along density layers, drifting southward. Decades later it may be part of the deep Atlantic, still carrying that old signature, now slightly altered by mixing and by the chemistry of life and decay in the interior ocean. Centuries later, some of that water may reach the Southern Ocean, where winds and currents can haul deep water back toward the surface. When it finally resurfaces and exchanges gases with the atmosphere again, it is as if the ocean has opened a time capsule.

The loop also helps explain a puzzle that otherwise seems contradictory: the ocean can keep absorbing heat even when surface warming temporarily slows. The atmosphere and the uppermost ocean are in constant conversation. If winds, currents, and patterns of variability shift the way heat is transported downward, the surface can cool relative to the long-term trend while the deeper layers warm more. This is not the planet changing its mind about physics; it is the planet redistributing energy within a system that has many moving parts. The overturning circulation is one of the main pathways for that redistribution, especially for moving surface anomalies into the interior and for returning deep anomalies to the surface on longer timescales.

The human story of how this circulation entered the public imagination is tied to a scientist with a gift for metaphor and a willingness to take the long view. Wallace Broecker, writing in 1991, popularized the image of a "great ocean conveyor belt." The phrase stuck because it turned a complex, three-dimensional circulation into a picture anyone could hold: a loop carrying warm water north and cold water south. Broecker's broader warning was that the climate system has "moods" and can change abruptly. The ocean's overturning, by moving freshwater and heat around, can participate in those mood swings. The conveyor belt metaphor has limits—the ocean is not a single belt with a single speed—but it did something essential. It gave the public a handle on the idea that the ocean is not just a background for storms but an active engine with global reach.

A dramatic historical episode often invoked in this context is the Younger Dryas, a rapid return to near-glacial conditions in parts of the North Atlantic region about 12,900 years ago. One hypothesis links it to a surge of freshwater from melting ice sheets that may have disrupted North Atlantic deep water formation, weakening the overturning circulation and reducing northward heat transport. The details remain debated, and the Earth system then was not the Earth system now. Yet the story endures because it illustrates a principle: when you interfere with the delicate balance of temperature and salinity that allows surface water to sink, you can alter the loop that redistributes heat. Freshwater is not just water; in this context it is a density-altering agent with the power to change the ocean's vertical exchange.

Modern concerns about AMOC are not that it will simply stop overnight like a stalled engine, but that it may weaken or shift, with regional consequences. Freshening from Greenland melt, changes in rainfall patterns, and warming of surface waters can all influence density in the formation regions. Even without invoking catastrophe, a gradual weakening would matter: it could change sea level along Atlantic coastlines, shift storm tracks, alter marine ecosystems, and rearrange where heat is stored. The fact that these changes play out amid natural variability makes them hard to detect cleanly in short records, which is one reason the ocean's memory can be politically frustrating. The system we most need to monitor is also the system that refuses to give quick, unambiguous answers.

The thousand-year loop also connects directly to sea level, a link that is sometimes underappreciated because it lacks the immediacy of a heat wave. As the ocean takes up heat, it expands. That expansion is not uniform because the ocean warms unevenly and because currents redistribute heat. Deep-ocean warming contributes to sea level rise on long timescales. Even if greenhouse gas concentrations were stabilized and surface temperatures stopped climbing quickly, the deep ocean would keep adjusting. Dana Ehlert and Kirsten Zickfeld (2018) explored how thermal expansion can remain effectively locked in for centuries to millennia, even under scenarios where carbon dioxide is later removed from the atmosphere. The implication is sobering: the ocean's slow loop does not just delay warming; it delays *commitment*, leaving a tail of consequences that stretches far beyond the period of emissions.

There is a temptation, when confronted with thousand-year timescales, to mentally file them under "not my problem." But the loop is not a distant abstraction. It is already shaping lived experience. When marine heatwaves persist, they can be sustained not only by local weather but by ocean currents delivering warm water and by a warm upper ocean that resists cooling. Coral bleaching is a surface expression, but the background state—how much heat the upper layers have stored—is set by longer-term uptake and circulation. Fisheries respond to shifts in temperature and oxygen that can be connected to changes in ventilation and stratification. Even the humidity of the air over continents is influenced by ocean temperatures that reflect both immediate weather and long-term ocean heat content. The loop touches daily life because the ocean is where the climate system's slow variables reside.

The ocean's memory also changes how one should interpret the word "irreversible." In everyday language, irreversible means permanent. In climate science, it often means something more precise: a change that will not be undone on timescales that matter for human societies, even if the original forcing is reduced. Deep-ocean warming and the sea level rise it drives through thermal expansion are classic examples. The loop that carries heat downward and brings it back up does not run on election cycles or infrastructure lifetimes. It runs on the physics of density and the geography of basins. That is why even strong mitigation can coexist with continued change for a time: not because mitigation fails, but because the ocean is still catching up with a past imbalance.

If all of this sounds bleak, it helps to remember that a system with memory is also a system with predictability, at least in its broad strokes. The ocean smooths the climate response, preventing some extremes that would otherwise occur in a faster-reacting system. It has, in a grim sense, bought time by absorbing heat that would have made the atmosphere even hotter already. Yet buying time is not the same as erasing debt. The ocean's loop is patient, and patience cuts both ways. It delays some harm while ensuring that what has been added does not simply vanish when attention shifts.

The most honest way to end is to return to the tactile world, because the ocean's long loop is easiest to grasp when it is tied to something one can imagine holding. Picture a glass sphere filled with water and dye. Warm the top gently and cool the bottom and you can set up slow motion within it: plumes sinking, layers forming, a reluctant mixing that takes far longer than you expect. Now enlarge the sphere to the size of a planet, replace dye with heat and dissolved gases, and add salt, ice, and winds. The surface will always be the part we watch, because it is where storms are born and where the air touches skin. The deep will keep its own counsel, storing the consequences of today's imbalance in water that will not see the sky again for centuries, carrying forward a memory that is physical, chemical, and profoundly hard to hurry.

4.3 The Gravity of Carbon

Carbon in the air feels weightless. It drifts, mixes, forgets where it came from. Yet the carbon dioxide molecule, CO_2, is constantly being tugged downward by processes that look, at first glance, almost too ordinary to matter: gas dissolving into liquid, plankton growing and dying, crumbs of organic matter sinking like dust through a sunbeam. The ocean does not merely *hold* carbon; it sorts it, transforms it, and in doing so gives carbon a kind of gravity.

Two broad mechanisms do most of the work. One is physical chemistry: the tendency of gases to dissolve more readily in cold water than warm. Oceanographers call this the *solubility pump*. The other is biological: the tendency of living things to take up carbon near the surface and export it downward as they are eaten, excreted, or die. This is the *biological pump*. Together they explain why the ocean has absorbed a large fraction of humanity's carbon emissions and why that absorption is neither unlimited nor permanent on human timescales.

The starting point is the air—sea interface, a boundary that is only a few molecules thick but spans most of the planet. Carbon dioxide crosses it by diffusion. If the partial pressure of CO_2 in the air is higher than the effective CO_2 pressure in surface water, the ocean takes up gas. If the surface water holds more, the ocean releases gas. Wind matters because it roughens the surface, creating waves and bubbles that increase the contact area and stir the top layer, keeping the boundary from saturating. Temperature matters because cold water can hold more dissolved gas. That is why the ocean's carbon sink has a geography: high-latitude oceans, cold and stormy, are often regions of strong uptake, while warm tropical waters can be closer to balance or even sources in some circumstances.

Solubility alone, however, would make the ocean a far more powerful sponge than it really is. The reason lies in what happens to CO_2 after it dissolves. In seawater, carbon dioxide is not just CO_2. Some of it remains as dissolved CO_2, but a great deal reacts with water to form carbonic acid, which then dissociates into bicarbonate and carbonate ions:

$$\mathrm{CO_2 + H_2O \rightleftharpoons H_2CO_3 \rightleftharpoons HCO_3^- + H^+ \rightleftharpoons CO_3^{2-} + 2H^+}.$$

This chemistry is why the ocean can store so much carbon in the first place: most of the dissolved inorganic carbon is held as bicarbonate, not as free CO_2. By converting incoming CO_2 into other forms, the surface ocean can keep taking up more gas than it otherwise would.

But this same chemistry creates resistance. Adding CO_2 shifts the balance toward more bicarbonate and dissolved CO_2 and away from carbonate, while increasing hydrogen ions, which lowers pH. The crucial subtlety is that the surface ocean does not respond linearly: each additional unit of CO_2 becomes harder to absorb without pushing the chemistry further from its previous state. This is commonly summarized by the *Revelle factor*, named after Roger Revelle, a measure of how sensitive surface-ocean CO_2 pressure is to changes in dissolved inorganic carbon. A high Revelle factor means the ocean surface *pushes back*: a small increase in dissolved carbon causes a relatively large increase in the CO_2 pressure of seawater, which weakens the gradient that drives further uptake from the atmosphere. The ocean can absorb a great deal, but it does not do so as a bottomless pit. It behaves more like a spring that stiffens as you compress it.

Revelle's role in this story is often misrepresented as a cautionary footnote, but it is better understood as a moment when the atmosphere and ocean were first treated, quantitatively, as a coupled system with limits. In 1957, Revelle and Hans Suess wrote about the exchange of CO_2 between air and sea and the implications for rising atmospheric carbon dioxide from fossil fuels. Their key insight was not that the ocean would ignore emissions, but that rapid *exchange* is not the same as rapid *removal* of the perturbation. The surface ocean can trade CO_2 with the atmosphere quickly, but the chemistry of seawater and the slow transport into the deep ocean constrain how much additional carbon can be stored without changing the system's state. That distinction remains one of the most important dinner-party clarifications in climate science: the ocean is not a simple drain.

The solubility pump works best where waters are cold and become dense enough to sink. When a surface parcel in the North Atlantic or around Antarctica cools, it can absorb more CO_2 from the air. If it then sinks as part of deep or intermediate water formation, it carries that dissolved carbon into the ocean interior, separating it from the atmosphere for decades to centuries. In this way the solubility pump piggybacks on the circulation described earlier: physics loads the carbon onto water, and circulation buries it. The carbon is not destroyed; it is relocated to a part of the ocean where it cannot quickly return to the air.

This relocation has a practical consequence. The contemporary ocean removes on the order of a few billion tonnes of carbon per year (expressed as gigatonnes of carbon, GtC), and cumulative uptake since the industrial era is on the order of hundreds of GtC. Those numbers are large enough to have noticeably slowed the rise of atmospheric CO_2. Without ocean uptake, the fraction of emitted carbon remaining in the air would be much larger, and the greenhouse forcing would be stronger. Yet the same numbers also hint at a limitation: emissions from fossil fuels and industry have been roughly an order of magnitude larger than the ocean sink in many recent years. The ocean helps, but it does not cancel.

Why does biology matter in a story that begins with chemistry? Because chemistry mainly governs how CO_2 partitions between air and surface water, while biology helps determine whether carbon stays near the surface or is exported to depth. The biological pump begins with photosynthesis. Phytoplankton, microscopic organisms floating in the sunlit upper ocean, take up dissolved CO_2 (mostly in the form of bicarbonate) and build organic molecules. They are eaten by zooplankton, which are eaten by larger creatures, and so on. Much of the carbon is respired back to CO_2 in the surface layer, returning quickly to the pool that can exchange with the atmosphere. But not all of it. A fraction becomes sinking particles: dead cells, fecal pellets, discarded mucus houses, fragments of shells and tissues. Oceanographers call this slow snowfall of organic detritus *marine snow*.

The phrase sounds gentle, almost cozy, until one remembers what it implies: a constant rain of carbon from the surface to the deep. As marine snow falls, microbes and animals consume it, respiring carbon back to CO_2 at depth. This respiration uses oxygen and releases nutrients. The result is that deep waters accumulate dissolved inorganic carbon and become relatively low in oxygen compared with surface waters. This is not merely a biochemical curiosity. It shapes where fish can live, where oxygen minimum zones develop, and how the ocean will respond to warming and stratification. The biological pump is not just a carbon conveyor; it is a life-support system with trade-offs.

Only a small fraction of the organic carbon produced at the surface reaches the seafloor and is buried in sediments on long timescales. But "small fraction" can still be important because the ocean is so vast and because time is so long. Over geological time, burial of organic carbon and formation of carbonate sediments help regulate atmospheric CO_2. On the timescale of centuries to millennia, the more immediate relevance is that biological export and deep-ocean storage separate carbon from the atmosphere long enough to matter for climate.

A concrete, human-scale example of this hidden rain can be seen in the aftermath of a phytoplankton bloom. In many temperate and polar regions, spring brings increasing sunlight and, after winter mixing, a surface ocean enriched with nutrients. Phytoplankton multiply rapidly, turning the water green. Satellites can see these blooms from space; fishermen and sailors can sometimes see the color change with their own eyes. For a brief time, the surface ocean acts like a busy factory, converting dissolved carbon into living matter. Then the bloom collapses. Grazers catch up, nutrients are depleted, and much of the organic material begins its descent. Instruments lowered through the water column can detect spikes in sinking particle flux following blooms, and deep-sea sediment traps collect the residue like slow-motion rain gauges.

The emotional punch of this example is that it is both ordinary and planetary. A bloom is a seasonal event, easy to overlook amid storms and shipping lanes. Yet blooms collectively help maintain the chemical difference between surface and deep ocean, a difference that allows the ocean to store carbon in the interior without instantly returning it to the air. In the absence of the biological pump, much of the ocean's dissolved carbon would reside closer to the surface, and atmospheric CO_2 would likely be higher. Life, in this sense, helps stabilize the planetary thermostat, even as it is being stressed by rapid change.

The solubility pump and the biological pump are often described separately for clarity, but in reality they are entangled. Cold water that absorbs CO_2 can also be biologically productive when nutrients are available. Biology can lower surface CO_2 by drawing it down into organic matter, enhancing uptake from the atmosphere. Conversely, when organic matter is respired at depth, it increases dissolved CO_2 there, which can later be brought back to the surface by upwelling, potentially making some regions natural sources of CO_2 to the air. The Southern Ocean is a classic example of this duality: it is a major region of anthropogenic CO_2 uptake, yet it is also where deep, carbon-rich waters can ventilate back to the surface, complicating the net balance.

This complexity is one reason the ocean carbon sink varies from year to year. It depends not only on emissions and atmospheric concentration, but on winds, temperature patterns, and biological conditions. Large-scale climate modes can influence it. So can volcanic eruptions that temporarily cool the surface and change circulation. Even changes in air pollution can matter: sulfate aerosols cool the climate by reflecting sunlight, and reductions in such aerosols can reveal warming that had been masked, altering ocean heat uptake and potentially affecting circulation and carbon exchange. The ocean is a coupled system. Pull on one thread and several others shift.

The idea that the ocean might one day become a less effective sink, or even a source in some regions, is not a speculative horror story. It follows from well-understood physics and chemistry. Warming reduces gas solubility, so warmer surface waters tend to hold less CO_2 for the same atmospheric concentration. Stratification can reduce the supply of nutrients to the surface in some regions, potentially weakening biological export. Changes in circulation can alter where carbon-rich deep waters upwell. Meanwhile, as more CO_2 dissolves, the chemical buffer stiffens: the Revelle factor increases, meaning additional uptake becomes progressively harder without larger shifts in ocean chemistry.

This is where David Archer and Michael Eby's synthesis (2009) on the atmospheric lifetime of fossil fuel CO_2 becomes relevant. A common misunderstanding is that because the ocean exchanges carbon with the atmosphere, emissions will be quickly absorbed and forgotten. In reality, there is a rapid phase in which carbon is shared among the atmosphere, surface ocean, and land biosphere, followed by a long tail governed by slow processes: transport into the deep ocean, carbonate chemistry adjustments, and, over even longer times, reactions with seafloor carbonates and weathering of rocks on land. A portion of emitted CO_2 remains in the atmosphere for centuries to millennia, not because the ocean is idle but because the ocean is already chemically loaded and physically slow to rearrange.

It helps to be concrete about what "chemically loaded" means. The surface ocean contains a large pool of dissolved inorganic carbon, but only a small fraction is in the form of dissolved CO_2 that directly controls the air—sea gradient. Adding carbon shifts the speciation and reduces carbonate ion concentration. That reduction matters for organisms that build calcium carbonate shells, and it also matters for buffering capacity. The ocean can take up more carbon, but each additional increment has a larger impact on pH and carbonate chemistry. The ocean is performing the service of absorption by paying with its internal chemical balance.

There is another kind of gravity at work, more literal than metaphorical: sinking. Particles sink because they are denser than seawater. Dense water sinks because it is heavier. Both are acted on by gravity, and both remove carbon from the surface. Yet the ocean does not simply drop carbon into a vault. It moves it into a vast interior where it will be processed, mixed, and eventually returned. The timescale of that return is set by circulation, the thousand-year loop. The pumps give carbon downward motion; circulation gives it delay.

The idea of delay matters emotionally because it changes what "solution" feels like. It is tempting to imagine that if emissions were cut sharply, the ocean would immediately begin drawing down atmospheric CO_2 at an accelerating pace. Some drawdown would happen, but the system's response is constrained. H. Damon Matthews and Kirsten Zickfeld (2012) explored what happens in climate models after emissions are set to zero: temperatures do not instantly snap back, and carbon sinks evolve over time as the ocean and land continue to adjust. The ocean does not stop being a sink the moment emissions stop, but neither does it become an infinitely eager vacuum cleaner. It continues to take up carbon while also responding to warming, circulation changes, and its own chemistry.

A final example brings the abstraction down to something almost visible. In some parts of the ocean, especially near productive coastal upwelling zones, divers and fishermen encounter waters that feel "sharp" to life: low oxygen, sometimes corrosive to shells, carrying a smell of decay. These waters are not created by industrial emissions alone; they are shaped by natural upwelling of deep, respired carbon, by local productivity, and by circulation. But as atmospheric CO_2 rises and the ocean absorbs more, the baseline chemistry shifts. The same upwelled waters can become more acidic, more challenging for organisms with calcium carbonate structures. The gravity of carbon is not just a global average measured in petagrams; it shows up as a changed feel of seawater to the organisms that must build, breathe, and reproduce within it.

Carbon's downward tug, then, is both a gift and a constraint. The ocean has slowed atmospheric accumulation by absorbing a substantial share of what has been emitted. It has done so by dissolving gas into cold waters, by transforming CO_2 into bicarbonate and carbonate, and by letting life package carbon into particles that sink into darkness. Yet the same mechanisms impose limits and delays. The chemistry resists linear absorption. The circulation stores carbon on long timescales rather than eliminating it. The biological pump depends on ecosystems already stressed by warming, deoxygenation, and acidification.

To speak of carbon having gravity is to admit that the climate problem is not only about air. It is about the ocean's willingness to carry part of our burden, and about the price the ocean pays in return: altered chemistry, altered circulation, altered habitats. The surface may feel like the stage where climate change performs, but the deep ocean is where the props are stored, heavy with carbon that sank out of sunlight and will take a long time to rise again.

Chapter 5

The Breath of the Biosphere

Life does not merely exist within a climate; it actively constructs it through a ceaseless exchange of gases and energy. This chapter traverses the scales of biological influence, from the annual inhalation of Northern Hemisphere forests to the geological grinding of roots that locks carbon in stone, before confronting the volatile ancient carbon threatening to wake from the frozen north.

5.1 The Green Pulse

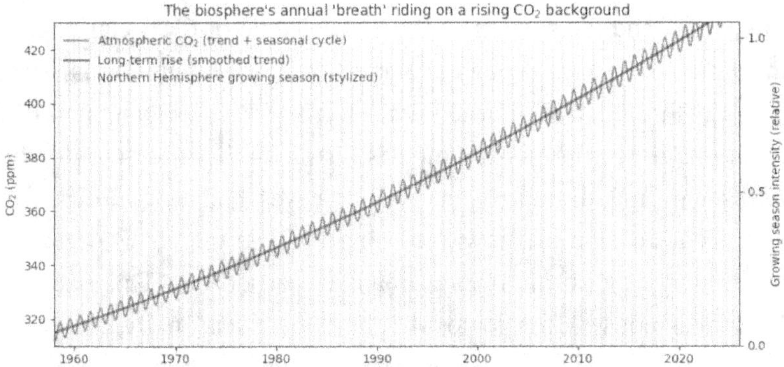

There is a way to watch the living world inhale and exhale without ever leaving the air. It appears as a jagged rhythm traced across decades: a line that climbs, falters, climbs again, and keeps climbing. At first glance it looks like the output of an industrial machine, all gears and momentum. Look more closely and you find something older and more intimate. Once each year, the biosphere takes a breath.

The record comes from a place that feels deliberately removed from ordinary life: the Mauna Loa Observatory on Hawai'i, perched high on the flank of a volcano. In 1958, Charles David Keeling began making careful measurements of atmospheric CO_2 there, choosing the site for its "clean" air and its distance from cities and forests that could muddy the signal. The story is often told as a triumph of patience, and it is. Keeling insisted on precision at a time when many scientists treated CO_2 as a vague background gas. He built instruments that could resolve changes of a fraction of a part per million and persuaded institutions to keep paying for the work year after year. That persistence produced a curve with two messages, layered like a palimpsest: the unmistakable upward march of human emissions, and the annual oscillation of life.

That oscillation is the fast carbon cycle, the one that moves on human timescales. Carbon is constantly being traded between air, vegetation, soils, and the surface ocean. A molecule of CO_2 can linger in the atmosphere long enough to matter for climate, but it is also swept up, transformed, and returned with startling speed. Plants pull CO_2 from the air during photosynthesis, using sunlight to stitch carbon into sugars and, ultimately, into wood, leaves, and roots. Respiration and decay reverse the flow, returning CO_2 to the air as organisms burn energy or as microbes dismantle dead tissue. This back-and-forth does not wait for geology. It turns over seasonally, daily, even hour by hour. The Keeling Curve simply makes the exchange visible.

The jagged pattern is sometimes described as a sawtooth, but it is closer to a slow sigh. CO_2 rises through autumn and winter, peaks in late spring, then falls through summer, reaching a minimum in early autumn. The timing surprises many people the first time they hear it. If the Northern Hemisphere greens up in spring, why does CO_2 keep climbing into May? The reason is that the atmosphere is the ledger for processes spread across continents. Through winter, photosynthesis is muted across the Northern Hemisphere, while respiration and decomposition continue. Forest floors do not go inert when leaves fall; microbes keep working whenever temperatures allow, and plants themselves respire in darkness. By the time spring warmth finally tips ecosystems into net uptake, the air has accumulated months of carbon release. The drawdown then accelerates through summer as leaves and needles act as biochemical machinery, pulling carbon from the air faster than it is returned. Only after the growing season wanes does respiration regain the upper hand.

A crucial detail is geographical: the Northern Hemisphere dominates the global seasonal signal because it has more land and more vegetation than the Southern Hemisphere. Most people live north of the equator, and so do most forests, farms, and soils that experience strong seasons. The Southern Hemisphere has plenty of biology, but it has a larger fraction of ocean and less land area at mid- and high latitudes where seasons are starkest. The consequence is that the global atmosphere inherits a Northern Hemisphere accent. Even a measurement station in the mid-Pacific cannot escape it. Air mixes around the planet, and the seasonal pulse propagates outward like a tide.

It helps to picture the atmosphere not as a static blanket but as a continuously stirred reservoir. Winds and weather systems churn air masses, spreading the imprint of continental seasons. In winter, cold, dark months limit plant uptake across Canada, Siberia, and northern

Eurasia. CO_2 accumulates. In summer, those same regions become a sprawling photosynthetic engine, and CO_2 is drawn down. One can almost imagine the boreal forests as a set of green bellows, vast enough that their collective metabolism shows up in the chemistry of the sky.

That does not mean the biosphere is *saving* us in the simple sense that trees are sometimes invoked. The seasonal cycle is mostly a redistribution of carbon, not a one-way disposal. Much of what is taken up in summer is returned in winter. A forest can be a net sink over years if it is growing or if carbon is being stored in soils and wood, but the annual inhale-exhale is primarily a pulse superimposed on a longer story. The Keeling Curve's two messages matter because they are easy to confuse. The oscillation can look large, even dramatic, and it is tempting to interpret the summer drawdown as a sign that nature will automatically cancel human emissions. The long-term trend is the rebuttal: the entire breathing pattern is riding an escalator.

Keeling's measurements were also an early lesson in how hard it is to keep a "clean" signal. Mauna Loa is remote, but it sits on a volcano that can emit CO_2, and the observatory sometimes experiences local influences from upslope and downslope winds. The solution was not to pretend the air was always pristine, but to treat the data with care: identify background conditions, filter out contaminated intervals, and build a record that reflects the broader atmosphere rather than local quirks. That methodological seriousness is part of why the curve became iconic. It was not merely a graph; it was an argument that the air itself can be monitored with the same rigor as blood chemistry.

Once you accept that the planet is breathing, a richer question follows: what, exactly, is doing the breathing? Leaves are the obvious answer, but the mechanism involves whole ecosystems. Croplands contribute, especially as agriculture expands and intensifies, but much of the seasonal amplitude is tied to forests and high-latitude land. Boreal regions are especially influential because they cover enormous areas and experience extreme seasonal contrast: long summer days with intense photosynthesis, followed by winters where uptake collapses but respiration never entirely stops. Even the timing carries ecological information. The CO_2 maximum tends to occur in late spring, while peak vegetation greenness often arrives later in summer. Photosynthesis ramps up as leaves unfold, but the atmosphere responds to the balance between uptake and release, and that balance is shaped by temperature, moisture, and the biological calendar of growth.

This is where the fast carbon cycle becomes more than a curiosity. Its seasonal rhythm is a diagnostic tool. Changes in the amplitude or timing of the CO_2 oscillation can signal shifts in ecosystems: longer growing seasons, altered plant productivity, changing patterns of drought stress, wildfire, insect outbreaks, or land-use change. Over recent decades, many researchers have pointed out that the seasonal exchange in the Northern Hemisphere appears to have intensified, suggesting that the biosphere's annual inhale-exhale has become larger. The reasons are still debated in detail: warming can lengthen growing seasons and speed decomposition; rising CO_2 can fertilize photosynthesis under some conditions; nitrogen deposition can alter productivity; forest regrowth can increase biomass; and climate extremes can abruptly reduce uptake. The atmosphere integrates all of it, which is both its power and its frustration. A single curve can tell you that something is changing, but not which lever is being pulled.

There is another layer of complexity hiding in plain sight: trees are not just chemical agents; they are physical objects in the energy balance of the planet. When a landscape shifts from pale grass or snow to dark evergreen canopy, it absorbs more sunlight. This is the albedo effect: albedo is a measure of reflectivity, and darker surfaces have lower albedo, meaning they absorb more incoming solar energy. In high latitudes, where snow can act like a mirror, replacing open land with dark forest canopies can increase warming even if the forest stores carbon. The climate effect of a forest is therefore the sum of at least two influences moving in opposite directions: CO_2 removal tends to cool by weakening the greenhouse effect, while surface darkening tends to warm by increasing absorbed sunlight. People sometimes ask whether planting trees is "good" or "bad" for climate as if the answer is a moral verdict. The physics refuses that simplicity. Location matters. Species and canopy structure matter. Whether the land would otherwise be snow-covered for months matters. A forest is a thermal actor as well as a carbon reservoir.

The seasonal CO_2 cycle also highlights an uncomfortable asymmetry between speed and permanence. Photosynthesis is fast. A spring flush of leaves can pull measurable CO_2 from the air in weeks. But storing that carbon in a way that matters for climate requires keeping it out of the atmosphere for decades to centuries. A leaf that falls in autumn is not long-term storage; it is carbon in transit. Even wood is not permanent if it burns, decays, or is harvested and oxidized. The breathing of the biosphere is therefore both reassuring and precarious: reassuring because it demonstrates that the living world is still metabolically vigorous on a planetary scale, precarious because metabolism can reverse.

One of the most elegant ways scientists separate these processes is by looking beyond CO_2 itself. The atmosphere carries tracers that act like fingerprints. Oxygen, for example, is produced by photosynthesis and consumed by respiration and combustion. If CO_2 rises while oxygen falls in a certain proportion, that points toward burning fossil fuels. Carbon isotopes provide another clue: plants preferentially take up lighter carbon, and fossil fuels have characteristic isotopic signatures. There is even a trace gas called carbonyl sulfide that plants absorb alongside CO_2, offering a window into photosynthetic activity. These additional measurements do not replace the Keeling Curve; they deepen it, turning a single line into a set of cross-checks about where carbon is coming from and where it is going. The result is a more intimate portrait of Earth's metabolism: not just that the planet breathes, but how the breath is partitioned among land, ocean, and industry.

The seasonal pulse can also be felt, in a small way, at the scale of ordinary experience. Walk through a forest on a bright summer day and you are surrounded by countless stomata, tiny pores on leaves that open to take in CO_2 and release water vapor. In aggregate, those pores influence humidity, cloud formation, and local temperatures. The air under a canopy can feel cooler not only because of shade, but because evaporation carries heat away. That local cooling sits alongside the global warming implication of the canopy's low albedo. Biology is not merely a passenger in climate; it is part of the machinery, sometimes pushing in opposite directions at once.

And then there is the human layer. Keeling's curve became a symbol not because it was the first hint that humans could alter the atmosphere, but because it rendered the alteration visible in a language everyone understands: a rising line. Yet the curve's annual wiggle is just as profound. It is the signature of fields, forests, and soils operating as a coordinated system across continents, without any central control, guided by sunlight and seasonality. The same atmosphere that records our emissions also records the quiet labor of photosynthesis, repeated each year with astonishing regularity.

If the long-term rise is the modern story of carbon, the seasonal oscillation is the older story of life's bargaining with air and light. It tells us that the biosphere is not a static backdrop but a dynamic, responsive partner in the planet's energy balance. Each spring, the Northern Hemisphere's vegetation begins to pull the CO_2 dial back, and each autumn it releases its grip. The motion is rhythmic enough to be predicted, yet sensitive enough to reveal change. The curve is therefore both heartbeat and fever chart: a steady pulse, and a sign that the patient is living through new conditions.

5.2 Rooting for Stability

The annual rise and fall of atmospheric CO_2 has the tempo of a living planet: quick enough to notice within a year, familiar enough to feel almost like weather. Yet Earth's temperature has also been kept within habitable bounds over spans of time so long that continents wander and species come and go. That kind of steadiness cannot be explained by a seasonal pulse. It requires a slower regulator, one that works like a thermostat with a very long delay: when the planet warms, the regulator strengthens a sink for carbon; when the planet cools, the sink weakens, allowing carbon to accumulate again. The surprising twist is that biology does not merely respond to this thermostat. In many places, biology helps turn the dial.

Start with a deceptively ordinary process: rock turning into dirt. If you pick up a handful of soil, you are holding the end product of a long chemical negotiation between minerals, water, air, and life. The technical term is *weathering*: the breakdown of rocks at Earth's surface. Weathering can be mechanical, like frost shattering stone, but the climate-relevant part is chemical. When certain rocks dissolve, they consume CO_2 in reactions that ultimately lock carbon away in dissolved bicarbonate and, much later, in carbonate minerals on the seafloor. The rocks most important for this are silicates, the planet's commonest crustal minerals: feldspars, pyroxenes, olivine, and their relatives. A basalt cliff, a granite boulder, the grains in a sandy riverbed are all dominated by silicates.

The chemistry sounds forbidding, but the basic idea is simple. CO_2

dissolves in rainwater to form a weak acid, carbonic acid. That slightly acidic water percolates through soils and cracks in rock, reacting with silicate minerals. In doing so it turns CO_2 from a gas into dissolved ions, especially bicarbonate (think of the fizz in mineral water, but stabilized). Rivers then carry these dissolved products to the ocean, where marine organisms and inorganic processes can precipitate carbonates, storing the carbon in rocks for millions of years. In shorthand: silicate weathering is a pipeline from atmospheric CO_2 to limestone.

The thermostat logic emerges from temperature and water. Warmer conditions tend to speed chemical reactions. A warmer world also tends, on average, to intensify rainfall and runoff, which flushes fresh water through soils and exposes more mineral surfaces to reaction. More warmth and water therefore tend to increase silicate weathering, which removes more CO_2 from the air, which weakens the greenhouse effect, which cools the planet. Cooling has the opposite effect, slowing weathering and allowing volcanic CO_2 and other sources to rebuild atmospheric levels. In 1981, James Walker, Paul Hays, and Jim Kasting gave a particularly clear formulation of this negative feedback, and it has been a cornerstone of thinking about long-term climate stability ever since.

But a thermostat is only as effective as the sensor and the wiring, and on Earth the wiring runs through roots.

To see why, it helps to look closely at what a plant is doing when it grows. Leaves pull CO_2 from the atmosphere, as the annual cycle makes vividly clear. Some of that carbon is sent underground, not as inert storage but as investment. Plants leak sugars and other compounds into the soil through their roots. They build fine root hairs to probe microscopic crevices. They trade carbon to fungi in exchange for nutrients. They alter the chemistry of the soil around them in ways that would be hard to believe if you had not seen the evidence:

the pH can shift; organic acids appear; carbon dioxide levels in the soil air can rise far above those in the atmosphere because roots and microbes respire in confined spaces. This zone, where roots and microbes crowd around mineral surfaces, has its own name: the *rhizosphere*. It is a thin sheath of altered chemistry around each root, but multiplied across a forest, it becomes a planetary-scale chemical factory.

Here is the essential link to weathering: acidic conditions and high CO_2 in soils make water more chemically aggressive toward silicate minerals. Add to that the physical action of roots and fungal filaments exploring cracks and mineral boundaries, and you get a biologically intensified weathering environment. The phrase used by geochemists is *biotic enhancement of weathering*. It is not a single mechanism but a suite of effects: respiration that enriches soil CO_2, organic acids that chelate metal ions and help dissolve minerals, microbial films that change local saturation states, and nutrient uptake that prevents dissolved products from building up and stalling reactions. Even the mere act of holding water in soils for longer periods, rather than letting it run off quickly, can increase contact time with minerals.

This is more than a modern curiosity. It reshapes the deep history of climate. For most of Earth's history, land was barren rock and microbial crusts. Then plants arrived, first small and shallow-rooted, later taller and more ambitious. Over the Devonian period, roughly 400 million years ago, vascular plants evolved the plumbing and structural support that allowed them to grow upright, spread across continents, and send roots deeper into soils. Robert Berner's GEOCARB model, developed in the late twentieth century, explored how such biological innovations could have shifted atmospheric CO_2 over hundreds of millions of years by strengthening weathering and organic carbon burial. The storyline implied by those calculations is dramatic: the rise of forests did not merely decorate the continents with green. It may have helped draw down CO_2 enough to cool the climate and reshape habitats on a global scale.

It is difficult to overstate how strange that is as an idea. Trees are not usually imagined as agents capable of changing the chemistry of mountains. Yet the evidence, accumulating from multiple angles, supports the general picture. Modern experiments help make it tangible. In controlled studies, researchers have grown trees with different mycorrhizal partnerships on crushed rock and measured how quickly key elements, like calcium, are released from silicate minerals. Work by Joe Quirk and David Beerling (2012), for example, emphasized that certain tree-fungus combinations can intensify mineral weathering, leaving microscopic alteration features on mineral surfaces and increasing the release of weathering products compared with other symbioses. The details matter: biology is not a universal accelerator in the same way everywhere, and some experimental designs reveal stronger effects from nutrient uptake and the removal of dissolved products than from CO_2 itself. The broader point is that the ability of plants to allocate carbon belowground, and to recruit microbial and fungal allies, can measurably change how fast rocks dissolve.

A concrete image helps. Picture a fresh basalt flow, black and glassy when it first cools. Basalt is rich in calcium and magnesium silicates, and it weathers relatively quickly compared with many other rocks. Left alone, rainwater will slowly leach it, and lichens may colonize its surface. Now add a forest. As roots penetrate cracks, the rhizosphere becomes a patchwork of tiny chemical reactors. Fungal hyphae, hair-thin filaments, extend far beyond the root itself, pressing into pores that roots cannot reach. They deliver acids and enzymes, and they harvest nutrients liberated by weathering. In return, the plant feeds them carbon. The rock becomes, in a sense, part of the plant's extended metabolism.

This biological intensification has a climatic consequence precisely because it is coupled to temperature and water. Warm, wet conditions tend to boost biological activity as well as chemistry. That creates the possibility of a feedback loop with two reinforcing links: a warmer climate speeds chemical weathering directly and also tends to support more vigorous ecosystems (up to limits imposed by drought or heat stress), which further enhances weathering. Over long timescales, the net effect is stabilizing because it removes CO_2 more efficiently when climate is warm. But stabilizing does not mean gentle. The thermostat can overshoot and undershoot; the response is delayed by the time it takes for mountains to erode, soils to develop, and carbonates to accumulate in sediments. The planet's long-term stability is not a static equilibrium but a wandering balance that has, at times, permitted snowball glaciations and hothouse climates.

There is also a human angle, because once you understand the mechanism, a provocative question appears: can people deliberately accelerate this slow thermostat in a controlled way? The idea has a name that sounds almost too literal: *enhanced rock weathering*. The proposal is to crush silicate rocks such as basalt into fine grains and spread them on land, particularly on farms, where they can react faster with CO_2-rich soil water. Crushing increases surface area dramatically, like the difference between a sugar cube and powdered sugar. More surface area means faster reaction rates. If the dissolved bicarbonate ultimately reaches the ocean and becomes carbonate, the CO_2 removal can be long-lasting.

The concept is old enough to have a historical flavor. In the nineteenth century, long before climate change was a public concern, chemists and agronomists debated how minerals replenish soil nutrients. Basalt and other rock powders were sometimes promoted as soil amendments. The modern twist is to track carbon as well as fertility. Today, field trials at watershed scales have reported measurable chemical signals consistent with increased weathering: changes in alkalinity and dissolved ions in drainage waters that imply additional CO_2 consumption. Yet the ranges are wide, and the practical difficulties are sobering. Where should the rock be quarried? How much energy is required to crush and transport it? How do you verify, with high confidence, that extra dissolved bicarbonate corresponds to net atmospheric CO_2 removal rather than simply shifting carbon among reservoirs? What happens to soil pH, trace metals, and ecosystems? The thermostat may be slow, but the attempt to tinker with it quickly runs into the complexity of real landscapes.

That complexity brings us back to the deeper lesson: biology does not operate in isolation from physics and chemistry. A forest is a chemical engine for carbon, but it is also a physical modification of the surface, changing albedo, roughness, and the water cycle. In the weathering thermostat, biology's role is more subterranean and less visible than the darkening of a canopy, but it is no less real. Roots and microbes change the chemical *pace* at which the crust can draw down CO_2. They also change the *place*: weathering is not uniformly distributed across continents. It is concentrated where fresh rock is exposed and where water flows through reactive minerals, often in mountainous regions and in the soils downslope from them. Life's contribution can therefore be geographically specific. A lush tropical hillside with deep soils may be a powerful weathering environment, but so might a temperate watershed with high runoff and reactive lithology. The map of climate regulation is not simply a map of forests; it is a map of geology braided with biology.

There is a paradox here that is worth sitting with. The fast carbon cycle feels like the realm of the living: leaves, seasons, the visible greening of spring. The slow carbon cycle feels geological: mountains, rivers, seafloor sediments. Root-mediated weathering dissolves that boundary. It makes the slow cycle partly biological, and it makes biology a participant in a planetary feedback that operates over millions of years. The planet's thermostat is not a single device. It is an emergent property of interacting systems: volcanic degassing that supplies CO_2, weathering that consumes it, oceans that store and circulate it, and life that both responds to and reshapes the chemical pathways.

A final, grounding example comes from a common experience in mountain country. After a heavy rain, streams often run milky or tea-colored, carrying fine particles and dissolved substances eroded from soils and rock. That water is not just scenery. It is the visible transport arm of the weathering thermostat, carrying ions that were, in a chemical sense, paid for by CO_2. Every river is a conveyor belt connecting air to stone to sea. The next time a stream swells and turns cloudy, it can be seen as more than weather: a momentary surge in the slow, collective work of transforming mountains into dissolved chemistry, aided by roots that have learned to treat rock as both foundation and food.

5.3 The Thawing Archive

A useful thermostat is not only about the mechanisms that cool. It is also about the mechanisms that can fail, or reverse direction, when conditions change too quickly. The Arctic is where that possibility feels least abstract, because the ground itself stores memory. Permafrost is not a material so much as a condition: soil or sediment that remains frozen for at least two consecutive years. In many places it has been frozen for thousands, even tens of thousands, of years. It is an archive in the literal sense that it preserves the past—dead roots, leaves, animal remains, ancient microbes, even the chemical traces of old climates. It is also an archive in the climatically dangerous sense that the preserved material is made of carbon, and carbon is eager to return to the air when thaw gives microbes permission to eat.

The scale is hard to grasp without numbers, and the numbers are unsettling. The northern permafrost region stores on the order of $\sim$ 1300 gigatonnes of organic carbon (often written as Pg C, petagrams of carbon), with substantial uncertainty and important deep deposits in river deltas and in ice-rich sediments known as Yedoma. That quantity is not a single, neatly measured stockpile; it is an estimate stitched together from field cores, laboratory analyses, and maps of soils that are imperfect even in accessible regions. The uncertainty is part of the story. An archive that cannot be fully inventoried is still an archive, and still capable of releasing what it contains.

What makes permafrost carbon different from the carbon in a living forest is not merely that it is old. It is that it has been physically locked away by temperature. In most ecosystems, plant litter falls to the ground and begins a race between burial and decomposition. Microbes compete with time: the longer organic matter remains warm and wet with oxygen available, the more completely it is converted back to CO_2. Cold slows microbes down. Freezing stops many processes almost entirely. Over millennia, repeated freeze—thaw cycles, waterlogged conditions, and slow burial can preserve organic matter that would otherwise have been respired. The Arctic has been running that preservation program since the last ice age and beyond.

Thaw changes the rules abruptly. Once frozen ground warms and ice melts, organic matter becomes accessible. Microbial decomposers awaken, multiply, and begin converting that carbon into gases. If oxygen is present—as in well-drained soils—the main product is CO_2. If the thawed ground is waterlogged and oxygen-poor, different microbial communities dominate and methane (CH_4) becomes a major product. Methane matters because molecule for molecule it is a much stronger greenhouse gas than CO_2 over the first few decades after it is emitted, even though it does not last as long in the atmosphere. The atmosphere is, in effect, more sensitive to a pulse of methane than to the same carbon released as CO_2 in the near term. That difference in potency is why permafrost is discussed not only as a carbon source, but as a potential amplifier of warming.

The amplification can be described with the language of feedback loops. A negative feedback damps change: more warmth triggers a response that cools. A positive feedback does the opposite: more warmth triggers a response that adds still more warmth. Permafrost thaw is a candidate for a positive feedback because warming increases thaw, thaw increases greenhouse gas release, and greenhouse gas release increases warming. The loop sounds tidy on paper. On the ground it is messy, because permafrost is not a uniform layer of frozen soil waiting to melt evenly like ice in a glass. It is a landscape feature that can fail gradually in some places and catastrophically in others.

One of the most vivid kinds of permafrost change is thermokarst, a word that combines the idea of thaw (thermo-) with the sinkholes and pits of karst landscapes formed by dissolving limestone. In ice-rich permafrost, the ground contains wedges, lenses, and layers of ice. When that ice melts, the soil collapses. Hillsides slump. Surface water reroutes itself. Ponds appear where none existed, or drain away where they once held water. From above, a thermokarst landscape can look like a quilt being pulled apart, with new hollows and wet patches developing in irregular patterns. The carbon consequences follow the water: collapse often creates saturated, oxygen-poor conditions that favor methane production. It also exposes fresh organic material, not only at the surface but at depth, by breaking open soils that were previously insulated and frozen.

A concrete example makes it easier to feel the mechanism. In parts of Alaska and Siberia, observers have documented lakes forming in newly collapsed ground, sometimes expanding rapidly as their warm water accelerates thaw along the edges. The lakes can bubble with methane, released from sediments below. If you walk on the ice of such a lake in winter, you can sometimes see trapped gas pockets frozen into the ice, like stacks of coins. Researchers have learned to measure these emissions with floating chambers, towers, and isotopic analyses that distinguish methane produced by microbes from methane leaking from deeper geological sources. The bubbling is not a metaphor. It is a direct transfer of ancient carbon from frozen soil to the modern atmosphere.

The history of how this became a scientific concern is also revealing. For much of the twentieth century, permafrost research was driven by practical needs: building roads, pipelines, and foundations that would not fail when the ground shifted. Engineering projects in Alaska, Canada, and Russia turned frozen ground into an object of intense scrutiny. The Trans-Alaska Pipeline, completed in the 1970s, is an emblem of that era: its elevated sections and carefully designed supports were responses to the simple fact that warm oil and warm air can destabilize frozen soil. In a sense, infrastructure forced attention onto permafrost long before climate models made it globally famous. The ground was already reminding people that "frozen" is not the same as "permanent."

As climate warming became unmistakable, the focus broadened from local stability to global feedback. Scientists began to treat permafrost not only as an engineering constraint but as a component of the carbon cycle. That shift required new kinds of measurements: mapping soil carbon stocks, monitoring thaw depth (the "active layer" that thaws each summer), tracking greenhouse gas fluxes across seasons, and building models that link physical thaw to microbial metabolism. The resulting picture is now clear enough to be worrying and nuanced enough to resist simple headlines.

One of the most important nuances is that the existence of a single, global permafrost "tipping point" is debated. The popular imagination likes thresholds: a line crossed, a switch flipped, a cascade that becomes unstoppable. Permafrost does have thresholds at local scales—a slope fails, a lake forms, a fire strips insulating vegetation and triggers deeper thaw. But when researchers synthesize across the entire Arctic, many conclude that the aggregate response is likely closer to quasi-linear with warming: more warming produces more thaw and more emissions, but without a single global cliff edge. That should not be comforting. A process can be quasi-linear and still profoundly consequential, because the feedback accumulates with every increment of warming. The absence of a single doomsday threshold does not imply safety; it implies a meter that keeps running.

A second nuance concerns timing and forms of carbon. CO_2 and methane behave differently in the atmosphere. Methane's near-term potency means that thaw-driven methane can act like a powerful shove, increasing warming quickly, while CO_2 acts like a long-term load added to the climate system. Permafrost emissions are therefore not just about how much carbon is released, but about the balance between gases and the timescales of their influence. That balance depends on hydrology: wet, oxygen-poor conditions tend to favor methane, while drier, oxygen-rich conditions favor CO_2. Climate change itself alters hydrology in contradictory ways, drying some regions and wetting others. Even within a single landscape, a hummock can be aerobic and a nearby hollow anaerobic, producing different gases from similar organic material.

Then there is the question of abrupt thaw. Many models represent permafrost thaw as a gradual deepening of the active layer: each summer thaws a little deeper, each winter refreezes, and over decades the perennially frozen layer retreats. That gradualism captures part of reality, but it can miss the drama of thermokarst and other collapse processes. Field studies have shown that abrupt thaw features can strongly amplify warming-driven CO_2 release relative to adjacent intact terrain. In at least one system, experimental warming in collapsed thermokarst features yielded about $5.5\times$ higher warming-induced CO_2 release than nearby intact ground. The number is not a universal multiplier; it is an illustration of a general principle: when ground structure fails, carbon that was protected by both temperature and physical isolation can become suddenly available.

This matters for prediction. Long-horizon modeling studies that extend centuries into the future, sometimes out to the year 2300, often find no self-perpetuating permafrost tipping in frameworks that include only gradual thaw. They also tend to caution that abrupt thaw processes are not fully represented, and that this omission can lead to underestimates of near-term emissions or of localized hotspots. The permafrost feedback is therefore not a single number but a range of plausible outcomes, and the width of that range is partly a measure of how heterogeneous the Arctic is.

The human stakes of that uncertainty are not confined to climate graphs. Thaw affects Arctic communities directly, warping roads and runways, tilting buildings, breaking water and sewage lines, and forcing costly repairs or relocations. When the ground shifts, it does not do so politely. It cracks and slumps, often during the short warm season when construction is hardest. These impacts turn permafrost from an abstract global concern into an immediate social and economic burden. That burden is unevenly distributed: communities that contributed least to global emissions often face some of the earliest and most expensive consequences.

Wildfire adds another layer. In boreal forests, fire is a natural part of the ecosystem, but warming has altered fire regimes in ways that can compound permafrost thaw. A severe fire can remove the insulating layer of moss and organic soil, exposing darker ground that absorbs more sunlight. It can also change vegetation, shifting from spruce to deciduous species or shrubs, with different effects on snow accumulation and shading. The ground then warms more in summer and freezes less in winter, deepening thaw. Fire is therefore both a carbon source in its own right and a facilitator of permafrost carbon release, intertwining fast and slow processes in a way that is difficult to untangle.

Despite all this, it would be a mistake to picture permafrost as a vault that suddenly empties. Much of the stored carbon will remain in place for centuries even under substantial warming, because some deposits are deep, some landscapes remain cold, and microbial decomposition is constrained by nutrients, moisture, and the physical structure of soils. The danger lies not in the idea that all permafrost carbon will be released, but in the reality that a fraction of a very large number is still large enough to matter, and that the release can come with an especially climate-effective blend of gases and timing.

There is also a subtle psychological trap. The phrase "feedback loop" can make permafrost sound like an external force acting upon society, a natural mechanism that will simply happen to us. But feedbacks are not independent actors. They are responses to conditions we create. The loop begins with warming, and warming is driven primarily by human greenhouse gas emissions. Permafrost emissions then add to that warming, raising the cost of delay and the value of restraint. In practical terms, it means that reducing emissions sooner can limit the amount of permafrost thaw that occurs, which limits additional emissions from permafrost, which limits additional warming. A positive feedback can be weakened by reducing the initial push.

It also means that precise predictions are less important than direction. Even if the exact magnitude of the permafrost feedback remains uncertain, the sign is not: thawed organic matter does not, on balance, remove greenhouse gases from the air. It adds them. The archive is not a library that stays quiet when the room warms. It is a pantry that begins to spoil.

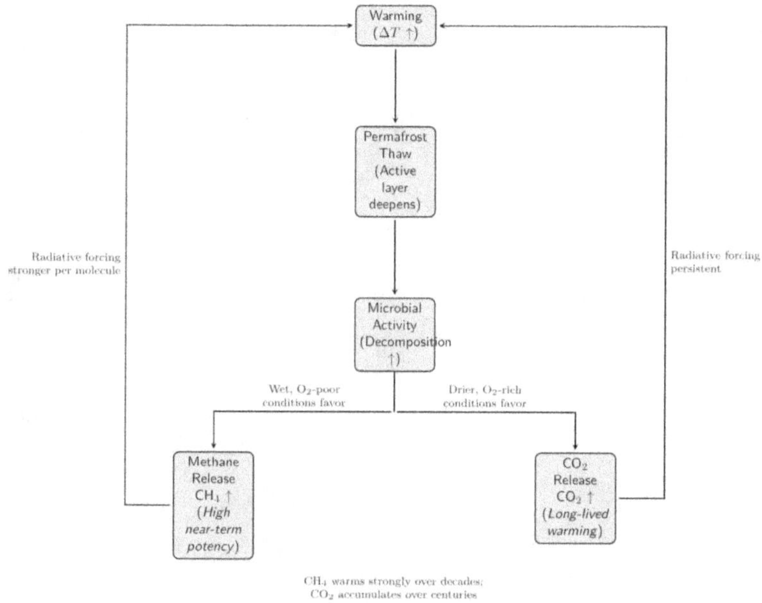

The loop drawn above is simple enough to fit on a page, but it points to something that is emotionally difficult to accept: time is stored in places. A frozen peat layer is not merely soil; it is a ledger of ancient summers, and those summers left behind carbon that was never fully returned to the air. When thaw begins, the ledger does not stay closed. It is read out as gas, molecule by molecule, by organisms too small to see. On a calm day in the far north, a faint odor from a thaw pond can be the smell of that reading: organic matter returning to chemistry, and chemistry returning to climate.

Chapter 6

The Cloudy Crystal Ball

Of all the variables in the climate equation, none are as fickle or powerful as clouds, which simultaneously shield Earth from the sun and trap its escaping heat. We delve into the physics of this atmospheric tug-of-war, exploring how shifting cloud altitudes and microscopic pollution particles act as the unpredictable wildcards in our planetary future.

6.1 Mirrors and Blankets

A cloud can be the most ordinary thing in the world—the kind you stop noticing on the drive to work—and also one of the sharpest tools in Earth's climate kit. The trick is that clouds do two different jobs at once. They meddle with *incoming* sunlight and with *outgoing* heat. Those two streams of energy travel in different forms, and clouds do not treat them equally.

Sunlight arrives mostly as shortwave radiation: visible light and its near neighbours in the spectrum. Earth, warmed by that sunlight, emits longwave infrared radiation: heat trying to escape back to space. A cloud is made of countless tiny particles—liquid droplets, ice crystals, or both—and each particle scatters and absorbs radiation in its own way. Put enough of them together and a cloud becomes, depending on its type, a mirror held up to the Sun or a blanket thrown over the planet.

The split personality is easiest to grasp by comparing two familiar characters: low stratus (and its close cousin, stratocumulus) and high cirrus. Low stratus clouds are often thick, grey, and close to the surface. High cirrus clouds are thin, feathery streaks drawn high across the sky. They can coexist on the same day, even over the same place, but they push the climate system in opposite directions.

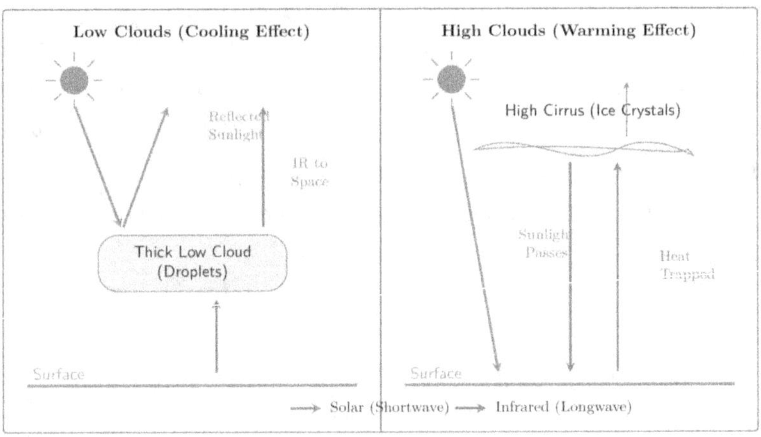

A useful way to keep the accounting straight is to imagine what happens at the *top of the atmosphere*, the notional surface where satellites measure the net flow of energy to and from space. Anything that sends more sunlight back out without letting extra heat escape is a cooling influence. Anything that reduces the amount of heat escaping without reflecting much extra sunlight is a warming influence. Clouds can do both simultaneously, and the balance depends on where they sit and how they are built.

Low stratus clouds tend to be bright. Stand beneath a solid marine layer at noon and the world looks evenly lit, as if someone has swapped the Sun for a giant lampshade. That brightness is not just a visual impression; it is physics. Low clouds are packed with liquid droplets, often tens of micrometres across, present in enormous numbers. For incoming sunlight, such droplets are excellent scatterers. A thick low cloud deck can reflect a large fraction of shortwave radiation straight back to space. Because the cloud top is close to the surface, it is relatively warm compared with the upper troposphere. Warm objects emit more infrared radiation than cold ones. So while the low cloud does absorb and re-emit infrared, it re-emits from a warm level and therefore still allows a substantial longwave flux to escape to space. The mirror effect is usually stronger than the blanket effect: low, thick clouds cool the planet on average.

High cirrus clouds reverse the emphasis. They are made mostly of ice crystals, often sparse enough that the cloud looks like a brushstroke. Sunlight can pass through with only modest reflection, especially compared with a low, liquid deck. But cirrus clouds live in colder air. The cloud top, which is the level from which much of the infrared radiation escapes, is frigid. A cold emitter sends less longwave radiation to space than a warm one. The cirrus acts like an elevated ceiling that forces Earth's heat to leave from a higher, colder altitude. The result is a reduction in outgoing infrared radiation: a warming influence. Cirrus can still reflect some sunlight, and the two effects can partially cancel, but the classic tendency of thin, high clouds is toward net warming.

This difference—warm low mirrors, cold high blankets—is the foundation. Yet the climate question is not only what today's clouds do, but how the balance changes as the world warms. A thermostat depends on feedback: a change in temperature causing a change in some process that either damps the original change (negative feedback) or amplifies it (positive feedback). Clouds are so unruly because they are not a single dial but a whole panel of levers: coverage, thickness, droplet size, crystal shape, altitude, lifetime, and the large-scale circulation patterns that herd them into place.

Start with a concrete scene. Imagine the summer subtropical eastern oceans—off California, Peru, Namibia—where cold upwelling water chills the air from below. The result is a stubborn, widespread stratocumulus deck that can persist for days, sometimes weeks, like a lid on the sea. These regions are small in area compared with the tropics as a whole, but they punch above their weight in sunlight reflected to space. If that low cloud cover thins or breaks up, even slightly, more solar energy enters the ocean, raising sea surface temperatures and encouraging further changes in the atmosphere above. This is one reason low clouds are treated with special respect in climate science: they sit in places where a small change can yield a large radiative consequence.

Now shift to the tropics, where warm, moist air rises in towering convective clouds. As those storms spread out near the top of the troposphere, their anvils become extensive layers of ice cloud. In satellite images, tropical convection often looks like a bright, cold shield. These high clouds are the signature of intense upward motion, and their area and altitude are tied to how deep the convection goes. If warming pushes the level at which air becomes stable upward, or changes the organization of storms, the height and extent of cirrus anvils can change too. A higher anvil top is colder; all else equal, that strengthens the blanket effect.

The atmosphere is not obliged to change "all else equal." High clouds can become thinner even as they rise. Low clouds can shrink in area but become brighter where they remain. The difficulty is that cloud fields are an emergent property: you cannot specify them simply by writing down temperature and humidity. They are the collective outcome of turbulent eddies, microphysics, and the broad conveyor belts of winds. That is why clouds are among the largest remaining uncertainties in the estimate of climate sensitivity: how much warming results from a given push to the energy balance.

A small historical anecdote shows how quickly the argument can turn from abstract to visceral. During the Second World War, the United States and Britain, worried about the effect of aircraft contrails on visibility and perhaps on weather, supported careful observations of condensation trails and high cloud. The science was primitive by modern standards, but the underlying curiosity was the same: thin lines of ice high in the sky might matter. Decades later, the point was driven home accidentally. In the days after September 11, 2001, commercial flights were grounded across the United States. That unusual pause offered an inadvertent experiment: with far fewer contrails, did surface temperatures change? Several analyses suggested that the daily temperature range increased slightly, consistent with the idea that high, thin clouds (including contrails that spread into cirrus-like veils) can reduce daytime heating a bit by reflecting sunlight while reducing nighttime cooling by trapping infrared, narrowing the day—night swing. The magnitude and interpretation have been debated, because weather varies for many reasons, but the episode became a cultural reference point: a reminder that delicate-looking high clouds can nudge the surface energy budget in detectable ways.

The story of low stratus contains a different kind of human imprint. Sailors in the eastern Pacific have long known the marine layer as a navigational mood: fog horns, damp decks, a dim horizon. For climate, the same layer is an enormous reflector. One of the simplest mental experiments is to ask what would happen if you could swap a patch of dark ocean for a patch of bright cloud. Ocean water absorbs most incoming sunlight; a thick low cloud can reflect a large fraction. Over an ocean, then, low clouds are like a sunshade that costs almost nothing in longwave terms because the cloud top is not dramatically colder than the sea beneath it. Over land or over snow, the arithmetic shifts. If the surface is already bright, adding cloud reflection buys less additional cooling, while the infrared trapping can still matter. That is why the same cloud type can have different net effects depending on the background.

Altitude is the cleanest divider between mirror and blanket, but thickness matters almost as much. A thick cloud is usually a better shortwave reflector than a thin one. Thickness is a measure of optical depth: how many opportunities photons have to be scattered or absorbed before they escape. Thick low clouds have large optical depth in the visible. Cirrus, by contrast, often has small optical depth. Yet even thin cirrus can have a surprisingly strong infrared influence because ice crystals interact efficiently with longwave radiation. The details depend on particle size and shape, which is why cirrus microphysics—how ice forms, grows, and falls out—has become a lively research frontier.

One can turn the mirror/blanket idea into a neat physical rule. Outgoing infrared radiation to space can be imagined as coming from an "effective emitting level": the altitude where the atmosphere becomes transparent enough for infrared photons to escape. Greenhouse gases raise that level by absorbing infrared and re-emitting it; clouds can do the same. If a cloud forms high up, it can become the effective emitting level for a wide band of infrared wavelengths. Because temperature generally decreases with height in the troposphere, moving the emitting level upward reduces the emitted power. This is the blanket. A low cloud does not move the emitting level upward by very much because it sits in comparatively warm air; it mainly alters the shortwave side. A high cloud shifts the emitting level into colder air and so can reduce outgoing infrared substantially even if it does not reflect much extra sunlight.

But rules of thumb become treacherous when they are mistaken for laws. There are cirrus that cool (for example, thicker, lower ice clouds that reflect strongly), and low clouds that warm (for example, thin stratus at night over a bright surface). The climate effect is always the difference between two big numbers: shortwave reflected and longwave trapped. In many places and seasons those numbers partly cancel, so modest changes in cloud properties can flip the sign.

This is where the “crystal ball” becomes cloudy in a deeper sense. Clouds are not passive curtains; they are symptoms of motion. Low stratus clouds in the subtropics are maintained by a particular atmospheric arrangement: air descending from higher altitudes warms and dries, placing a cap (a temperature inversion) above a cool, moist boundary layer. Turbulence below the inversion keeps moisture mixed upward, feeding cloud at the top of the boundary layer. If warming strengthens the inversion in some regions but weakens it in others, the low cloud response will differ. If the surface warms faster than the air aloft, mixing changes. If winds shift, upwelling changes, and the sea surface temperature pattern changes too. The low cloud deck is, in other words, a visible tracer of a complicated set of controls.

High cirrus is similarly tied to dynamics, but of a different flavour. Much cirrus is born from deep convection: rising moist air that cools and condenses, lofting water that freezes at high altitude. Another portion forms in situ in the cold upper troposphere where air is supersaturated with respect to ice. The coverage and thickness of cirrus depend on how much water vapour reaches those heights, how quickly ice crystals grow and fall, and how frequently air parcels are lifted. Warming increases the capacity of air to hold water vapour, a basic thermodynamic fact familiar from muggy summer days. That might suggest more high cloud and stronger longwave trapping. But warming also changes the vertical temperature profile and the efficiency with which convection exports moisture upward. Some theories suggest that as the tropics warm, anvils may rise but thin, partly offsetting the stronger blanket that greater height would imply. The atmosphere can pay one bill by issuing another credit.

One reason these subtleties matter is that the net cloud influence is already large in the present climate. If Earth had the same greenhouse gases and surface properties but no clouds, the planet would absorb more sunlight and also emit more infrared directly from the surface and lower atmosphere. Clouds, in aggregate, cool Earth in the shortwave more than they warm it in the longwave, producing a net cooling at the top of the atmosphere. Yet the *feedback*—how clouds change with warming—need not share the sign of the present-day effect. A system can be cooling overall but still become *less* cooling as it warms, which is a positive feedback. Imagine a mirror that gradually tarnishes as the room heats up: the room still contains a mirror, but it reflects less, so the heating accelerates.

A useful everyday analogy is the difference between a white T-shirt and a thin jacket on a cool evening. In bright sun, the white shirt reflects and keeps you cooler; at night, the jacket reduces heat loss and keeps you warmer. A low marine cloud deck is closer to the white shirt at noon; high cirrus is closer to the jacket at night. Earth wears both at once, and the climate question is which garment changes in which direction as the planet warms.

On some mornings in coastal California, the marine layer burns off by midday, revealing blue sky. The timing depends on subtle things: sea surface temperature, winds, the strength of the inversion, and how much the land heats up and draws air inland. In climate projections, a small systematic shift in the persistence of that marine layer—not the existence of clouds everywhere, but the frequency and extent of a particular regime—can change the planetary energy balance measurably. People feel this locally as heat, glare, and dryness; the planet feels it as a change in albedo.

On other days, especially in the tropics, you can watch towering cumulus grow into thunderstorms and then spread a high, thin veil of ice downwind. Late in the afternoon, the Sun may be dimmed only slightly, but the sky takes on a milky appearance that persists after sunset. That veil is doing its quiet work in the infrared, slowing the loss of heat to space. If such veils become more common, or sit higher, the blanket thickens.

The uneasy beauty of the cloud problem is that it refuses to be reduced to a single moral. “More clouds” does not mean “cooler,” nor does “fewer clouds” mean “warmer.” The crucial details are where the clouds are, how high they reach, and how opaque they are in the right part of the spectrum. Low, thick clouds are among the most effective mirrors Earth has; high, thin clouds are among its most effective blankets. The same sky can contain both, and the climate is the running total of their tug-of-war: sunlight turned away before it can warm the surface, and heat delayed on its way out to space, held up for a little longer under a ceiling of ice.

6.2 The Invisible Architects

From the ground, clouds look like pure water made visible. From the inside, a cloud is closer to a crowded city: countless collisions, tiny traffic jams of droplets, sudden clearings, and the slow settling of heavier particles through the bustle. The surprising part is that most of the architecture is drawn by things far too small to see. A cloud usually cannot begin as “just water vapour cooling down.” It needs a scaffold: microscopic particles that give water molecules a place to gather.

Those particles are *aerosols*—solid or liquid specks suspended in air. Some are natural: sea salt lofted by breaking waves, dust lifted from dry ground, sulphur compounds oxidized from gases emitted by plankton, smoke from lightning-ignited fires. Some are made, directly or indirectly, by human activity: soot from incomplete combustion, sulphate formed from sulphur dioxide released by burning coal and oil, nitrates and organics from industrial chemistry and traffic. The word "aerosol" can sound clinical, like something in a spray can. In climate, it means the floating grit of the atmosphere, and it matters because it decides how many cloud droplets there will be and how long they will survive.

The key concept is *cloud condensation nuclei* (CCN). Water molecules in air are restless; they bump into each other, briefly forming clusters that fall apart almost immediately. To make a stable droplet, the cluster must grow past a size where surface tension is no longer an overwhelming penalty. An aerosol particle lowers that barrier. If it is hygroscopic—happy to absorb water, as sea salt is—it acts like a magnet for moisture. Even less hygroscopic particles can do the job if the air is sufficiently supersaturated (meaning the humidity is above what would be stable over a flat water surface). In practice, the atmosphere rarely makes cloud droplets without CCN. Aerosols are the seeds; the cloud is the crop.

That simple fact lets the invisible world tug on the visible one. More CCN generally means more droplets. If you have the same amount of condensed water but it is divided among more droplets, each droplet is smaller. This shift in droplet size changes the way the cloud interacts with sunlight. A cloud made of many small droplets tends to be brighter than one made of fewer large droplets. The reason is geometric: for a given volume of water, many small spheres have more total surface area than a few large spheres. More surface area means more scattering of incoming solar radiation, and a higher *albedo* (reflectivity). This is often called the *Twomey effect*, after Sean Twomey, who articulated in the 1970s how aerosol-driven changes in droplet number could alter cloud brightness.

One can watch the Twomey effect in the modern sky if one knows where to look. Satellite images sometimes show thin, bright lines running across a marine cloud deck like scratches on photographic film. These are *ship tracks*. A ship's exhaust adds extra aerosol particles to the air above the ocean. If the background marine clouds are made of relatively few droplets, that injection can raise droplet number and brighten the cloud along the ship's path. The tracks can stretch for hundreds of kilometres, a visible signature of an invisible perturbation. It is hard to find a more literal demonstration that a handful of particles, released at the surface, can rewrite cloud optics far above.

Ship tracks also serve as a cautionary tale, because they do not always brighten clouds. Under some conditions, adding aerosols can leave cloud reflectivity nearly unchanged, or even reduce it. Mixed-phase clouds—those containing both supercooled liquid droplets and ice—can respond differently than purely liquid clouds. Meteorology matters: if the atmosphere is dry above the cloud, changes in droplet size can alter how readily the cloud entrains that dry air and evaporates. A "seed" that ought to make a cloud more reflective can, in a different setting, help a cloud thin or break. The moral is not that the microphysics is unknowable, but that it is conditional. Aerosols pull levers that are connected to other levers.

Brightness is only half the story. Aerosols can also affect cloud *lifetime* and *coverage* by influencing precipitation. In many warm clouds, raindrops begin when droplets collide and coalesce into larger drops. Larger initial droplets make this easier; a cloud with fewer, bigger droplets can drizzle more readily. If extra aerosols produce many smaller droplets, collisions become less efficient and drizzle can be suppressed. When drizzle weakens, cloud water can persist longer and the cloud can spread, reinforcing its reflective effect. This suite of ideas has often been called the "cloud lifetime effect," although researchers have learned to treat it with care: cloud systems are embedded in dynamics, and what looks like longer-lived cloud might be a change in circulation rather than microphysics alone. Still, the basic linkage—droplet size influences drizzle, which influences cloud persistence—is physically real and frequently observed.

At this point the story intersects uneasily with human history. The twentieth century was a massive aerosol experiment conducted without consent forms. Industrialization poured sulphur dioxide, soot, and other precursors into the lower atmosphere. In many regions, especially downwind of coal-burning power plants and heavy industry, haze became an everyday feature of the sky. That haze did two things relevant to climate. First, some aerosols scatter sunlight directly, cooling the surface by shading it. Second, aerosols altered clouds, often making them brighter and changing their behaviour. Both effects tended to oppose greenhouse warming, at least regionally and during the period of most intense pollution.

A concrete episode brings the entanglement into focus. In 1991, Mount Pinatubo in the Philippines erupted and lofted a vast plume of sulphur-rich gases into the stratosphere. There, the gases formed sulphate aerosols that spread globally. The result was a measurable cooling of the planet for about a year or two, on the order of a few tenths of a degree Celsius. Pinatubo was not a cloud story in the everyday tropospheric sense—the stratospheric particles were more like a thin veil than a weather cloud—but it became a vivid lesson that tiny particles, distributed widely, can move global temperature. It also sharpened scientific interest in how aerosol particles, whether from volcanoes or smokestacks, interact with radiation and with clouds, and how quickly the climate responds.

Tropospheric aerosol pollution is more patchy and more complicated than a stratospheric volcanic plume. It is subject to rainout, chemical transformation, and regional circulation. Yet the principle remains: aerosols can mask some greenhouse warming by increasing reflectivity. That masking is not a stable, benevolent buffer. It is an accidental side effect of dirty air, and it comes with a brutal health cost. When societies reduce sulphur emissions and soot for public health reasons—one of the great environmental successes of modern regulation—the atmosphere can become clearer, and some of that cooling mask is removed. The planet then reveals more of the warming that greenhouse gases were already committed to producing. There is a cruel asymmetry in timescales: many aerosols last days to a week in the lower atmosphere, while carbon dioxide lingers for centuries. Cleaning the air can change cloud brightness quickly; cleaning the carbon budget is slower.

To talk sensibly about these competing influences, climate scientists often use the language of *radiative forcing*. A forcing is a push to Earth's energy balance: a change in the net flow of energy at the top of the atmosphere, measured in watts per square metre. Greenhouse gases provide a positive forcing by reducing outgoing infrared. Reflective aerosols provide a negative forcing by sending sunlight back to space. But aerosols are slippery because they do not merely impose an instantaneous optical change. They can trigger *rapid adjustments* in clouds: shifts in cloud cover, droplet size, and atmospheric stability that happen without waiting for the oceans to warm. For this reason, modern assessments often speak of *effective radiative forcing* (ERF), which bundles the instantaneous effect together with these fast responses. It is a reminder that the atmosphere is not a static pane of glass; it reorganizes when perturbed.

If the goal were only to predict global-mean temperature, one might hope to average over the messy details. Unfortunately, aerosols do not distribute themselves uniformly, and clouds are not uniform either. Much of the aerosol burden lies in the Northern Hemisphere mid-latitudes and in industrial regions, or downwind of them. Many of the most cloud-sensitive regimes lie over oceans and coastal zones.

The overlap is imperfect. That mismatch matters because forcing patterns can shift winds and rainfall. A regional cooling from aerosols can alter sea surface temperatures, which alters atmospheric circulation, which alters clouds again. In other words, aerosols can influence climate not only by a direct radiative push but also by changing the patterns of heating that steer the atmosphere.

A sobering example is the history of “global dimming” and “global brightening.” Measurements of sunlight reaching the surface, made with pyranometers at weather stations, showed that in many places the surface received less solar radiation from roughly the 1950s to the 1980s. Later decades showed partial recovery. The causes are multiple—instrumental issues, cloud changes, aerosol changes—but the broad narrative is that mid-century pollution, particularly sulphate aerosols, likely contributed to dimmer skies, and later air-quality improvements contributed to brightening. That shift affected not only temperature but also evaporation, plant growth, and hydrology. Farmers and water managers experienced it as changes in drying rates and drought stress long before it became a neat line on a graph. The atmosphere’s microscopic inventory leaked into human-scale decisions.

The cloud response to aerosols is not always cooling. Black carbon (soot) absorbs sunlight efficiently. When soot is suspended in the air, it can warm the layer in which it resides while shading the surface below. That vertical pattern can stabilize the atmosphere, suppressing convection and potentially reducing low cloud in some cases, which would warm. When soot settles onto snow or ice, it darkens the surface, accelerating melt. Even within clouds, absorbing particles can heat droplets and encourage evaporation, a process sometimes called the semi-direct effect. Once again, the aerosol category hides opposites: sulphate tends to cool by scattering; soot can warm by absorbing, even while also affecting clouds.

The phrase “invisible architects” fits because the aerosols do not merely decorate clouds; they decide the building codes. How many droplets form, how quickly they grow, whether they collide into raindrops, whether a cloud deck persists into afternoon or breaks into blue patches—these outcomes depend sensitively on aerosol number and chemistry. And aerosol populations shift with human decisions: fuel choices, shipping regulations, wildfire management, agricultural burning, industrial scrubbers, urban planning.

Shipping provides an especially tangible connection between policy and clouds. When international rules tightened the sulphur content allowed in marine fuels in the early 2020s, ships emitted less sulphur dioxide, reducing sulphate aerosol formation over busy shipping lanes. That change was driven by public health and environmental concerns, especially near coasts. But it also altered a long-standing source of aerosol that had been brightening marine clouds in certain regions. The sky above shipping routes does not issue press releases, yet it responds. Changes in emissions can reduce the frequency or intensity of ship tracks, subtly shifting cloud reflectivity. The effect on global climate is small compared with greenhouse gases, but it is a clean illustration of how quickly aerosol-driven cloud changes can follow regulation.

The details of aerosol-cloud interaction remain among the most stubborn uncertainties in climate science, partly because the atmosphere is a noisy laboratory. It is difficult to compare “the same cloud” with and without aerosols when the winds, humidity, and temperature are always changing. Laboratory experiments help with microphysics, but laboratory clouds are miniature and simplified. Satellites give global coverage, but they infer droplet size and aerosol properties indirectly, and disentangling cause and effect is hard: do aerosols change clouds, or do meteorological conditions that favour clouds also concentrate aerosols? Aircraft campaigns provide exquisite measurements, but only for short periods and limited regions. Progress comes from stitching these approaches together, using the strengths of each to constrain the others.

One way to appreciate the difficulty is to return to the most basic claim: "more aerosols make clouds brighter." Even that is conditional. If the background cloud already has abundant droplets, adding more CCN may do little. If the cloud is mixed-phase, ice processes may dominate its evolution. If the cloud is drizzle-prone, suppressing drizzle might thicken it; if the environment is dry, smaller droplets might evaporate more readily and thin it. If aerosols absorb sunlight, they may warm and dry the layer, reducing cloud cover. The statement is not false; it is incomplete. The atmosphere does not hand out single-cause experiments.

And yet, despite the complexity, the stakes are clear. The planet's temperature responds to the net energy balance: how much sunlight is absorbed versus how much infrared escapes. Clouds help set both terms. Aerosols, by governing clouds' droplet populations and lifetimes, act as a hidden hand on that balance. For decades, industrial aerosols likely offset a nontrivial portion of greenhouse warming. That offset was never reliable, never evenly distributed, and never free. It came attached to smog-choked cities, acidified lakes, damaged lungs, and premature deaths.

There is an uneasy ethical lesson embedded in the physics. If tiny particles can cool the planet by making clouds brighter, then in principle one could try to engineer that cooling deliberately. The idea appears in proposals for marine cloud brightening: spraying sea-salt particles to increase droplet number in low marine clouds, enhancing their reflectivity. The scientific motivation is understandable: low clouds are powerful mirrors. But the very sensitivity that makes the idea tempting also makes it hazardous. Cloud systems are entwined with regional rainfall and circulation. A change intended to cool might shift monsoons or marine ecosystems, and the benefits and harms would not be shared evenly. The invisible architects can be hired, but they do not take instructions politely.

For now, the more immediate reality is less speculative and more intimate. Every time a diesel engine runs, every time a wildfire sends smoke across a continent, every time waves break into spindrift under strong winds, the atmosphere's inventory of seeds is altered. Most of those seeds will never become part of a cloud; they will be washed out by rain or settle to the surface. A small fraction will help form droplets that brighten a morning marine layer or thin a cloud enough to let drizzle fall. In those tiny acts of formation and loss, repeated trillions of times, the sky makes its decisions—and the planet's thermostat receives one more quiet adjustment.

6.3 Smoke and Mirrors

A wildfire announces itself in the most human of ways: the bite of smoke in the throat, the orange bruise of a sun seen through haze, ash collecting in the corners of window frames. Those are local experiences. From above, the same fire can look like a machine designed to meddle with weather. It manufactures aerosols, rewrites cloud microphysics, changes how sunlight reaches the ground, and in extreme cases punches material into the stratosphere, where it can linger for months. Fire is not only a response to climate; it becomes an actor inside the climate system, a feedback that can run hot in more ways than one.

Smoke is a complicated substance. It contains black carbon (soot), organic particles, and a soup of gases that can later condense into new particles as the plume ages. Some of those particles scatter sunlight and cool the surface beneath; some absorb sunlight and warm the layer of air they occupy. The same plume can cool one place by shading it and warm another by absorbing solar energy aloft. This duality is the "mirrors" part of the story: the atmosphere fills with tiny reflectors and absorbers whose net effect depends on their composition, altitude, and the brightness of the surface below.

The "smoke" part is equally literal, but it is also about deception. Smoke can make the sky look overcast while the ground remains dry. It can thicken clouds while suppressing rain. It can give a short-lived reprieve from heat at the surface while quietly strengthening the conditions that favour the next fire. Wildfire feedbacks are not a simple loop of warming leading to more burning leading to more warming. They are a web of competing processes, many of them fast, some of them delayed, and several of them capable of flipping sign depending on context.

One place to start is with sunlight. A thick smoke plume dims the day in a way that feels almost theatrical. During major wildfire episodes in western North America, people have eaten lunch under streetlights at noon; in parts of Australia during the Black Summer of 2019—2020, the sky turned an uncanny brown. The physics behind that mood lighting is straightforward: particles scatter and absorb shortwave radiation, reducing the solar energy that reaches the ground. Less sunlight at the surface can mean lower afternoon temperatures, slower evaporation, and in some cases a brief reduction in the intensity of the very fire producing the smoke. This is a negative feedback, at least locally and temporarily: fire makes smoke; smoke shades the surface; shading reduces heating.

But shading is only half the radiative ledger. When sunlight is intercepted by smoke, it does not vanish; it is redirected or absorbed. If the smoke is absorbing, especially if it contains a substantial soot fraction, it heats the air layer where it resides. Warming aloft can stabilize the atmosphere, strengthening an inversion that caps mixing. A stronger cap can trap pollutants and moisture near the surface, changing boundary-layer clouds. In some marine environments, absorbing aerosols above cloud decks can reduce cloudiness by warming and drying the air just above the cloud top, enhancing evaporation. That can reduce low-cloud cover and increase sunlight reaching the surface, a warming influence that runs counter to the direct shading effect. The net result depends on whether smoke sits above or within cloud layers, and on how reflective the underlying surface is.

Now bring in the microphysics of clouds. Smoke particles are aerosols, and aerosols can act as cloud condensation nuclei. In regions where clouds form in smoky air, droplet number can increase and droplet size can decrease, often making clouds brighter in sunlight. Brighter clouds mean more reflection and a cooling tendency. At the same time, smaller droplets can suppress drizzle by making collisions less efficient. In some cases, that keeps clouds around longer, extending their reflective influence. This is the classic aerosol—cloud interaction story, but with a combustible twist: the aerosol source is not industry or shipping but a landscape on fire.

The twist is that smoke does not always brighten clouds. Mixed-phase clouds, common at higher latitudes and in cooler seasons, can respond in ways that defeat simple intuition. The presence of ice changes how clouds precipitate, and the relationship between droplet size, glaciation, and reflectivity becomes more tangled. Observations from ship tracks have already shown that aerosol perturbations can yield weaker brightening in mixed-phase clouds, and sometimes even reduced albedo under certain meteorological conditions. Smoke operates under the same atmospheric laws. A plume that would whiten a warm, liquid marine cloud might have a muted or even opposite effect in a colder cloud containing ice. This matters because many fire-prone regions export smoke into mid-latitudes where mixed-phase clouds are common, especially over the Southern Ocean and high-latitude storm tracks.

Even when smoke does brighten clouds, the rainfall story can go in unsettling directions. Suppressing drizzle sounds like a recipe for more cloud and perhaps more cooling. But rainfall is also a delivery system that cleans the air and moistens the land. If smoke-induced microphysical changes reduce precipitation in some situations, downwind landscapes can dry further. Drier fuels ignite more easily, and fires spread faster. Here is a plausible positive feedback: warming and drying increase fire; fire produces smoke; smoke alters clouds in a way that reduces rainfall; reduced rainfall increases drying and fire risk. The same particles that brighten a cloud can, through the hydrological side effects, help set the stage for more burning.

The most dramatic smoke—cloud interaction is not about subtle changes in droplet size. It is about thunderstorms made by fire. In extreme events, the heat released by a wildfire creates a vigorous updraft that lofts smoke and moisture high into the troposphere. If conditions are right, the plume develops into a pyrocumulonimbus cloud—often shortened to *pyroCb*. It looks like a thunderstorm because it is one: a towering convective cloud with an anvil top, sometimes producing lightning, hail, and strong outflow winds. The difference is that its engine is not primarily solar heating of the land and moisture convergence, but the concentrated heat of combustion.

PyroCbs matter because they can transport smoke to extraordinary heights. Ordinary wildfire smoke is mostly confined to the lower troposphere, where it is removed by rain and mixing on timescales of days to a week. A pyroCb can inject smoke into the upper troposphere and even into the lower stratosphere. The stratosphere is dry, stable, and relatively isolated from the weather systems that scrub the lower atmosphere. Particles placed there can persist for months, spreading over continents and oceans. This is stratospheric injection not by a volcano, but by a firestorm.

A concrete example arrived in a year many people remember for other reasons: 2020. During the Australian Black Summer, several pyroCb events lofted smoke high enough that it was tracked around the globe. Satellite instruments observed aerosol layers in the stratosphere that behaved, in some respects, like a modest volcanic plume. The smoke altered the radiative balance and even affected stratospheric temperatures. Another striking sequence occurred in western North America in 2017 and 2018, when intense fires in British Columbia and the western United States produced pyroCb events with stratospheric smoke. For atmospheric scientists, these episodes were a kind of shock: not because pyroCbs were unknown, but because their frequency, intensity, and hemispheric reach underscored how fire weather can couple into climate-relevant altitudes.

Once smoke reaches the stratosphere, the "mirrors and blankets" logic changes character. At those heights, particles interact with sunlight strongly and can cool the surface by scattering light back to space, much as volcanic sulphate aerosols do. But smoke is not sulphate. It can be more absorbing, and that absorption can warm the stratospheric layer itself. Warming aloft can change circulation patterns and influence ozone chemistry. The smoke layer can also modify the temperatures that determine where and how high clouds form below, because the atmosphere is a connected column: heat absorbed in one layer shifts stability in adjacent layers. The result is not a single clean forcing but a set of adjustments.

The feedback story deepens when lightning enters. PyroCbs can produce lightning that ignites new fires, sometimes kilometres away from the original front. That creates a direct positive feedback: fire breeds storms; storms breed lightning; lightning breeds new fire. On the other hand, thunderstorms can also produce rain that helps extinguish fires. In practice, pyroCb precipitation is often spotty and can evaporate before reaching the ground if the lower air is dry. Some events generate "dry lightning" and strong downdrafts that spread embers, worsening the situation. The fire-made storm becomes a weather hazard in its own right.

All of this sits atop the slower, landscape-scale feedbacks that determine how flammable a region becomes over years and decades. Warming increases evaporation demand. Earlier snowmelt extends the dry season in many mountainous areas. Heat waves stress vegetation, and drought kills trees, increasing dead fuel loads. Insect outbreaks, encouraged by warmer winters in some regions, can leave forests full of tinder. Human land-use patterns create ignition sources and fragment ecosystems. When a large fire burns, it can alter albedo and surface roughness, changing how the land absorbs sunlight and exchanges heat with the atmosphere. It can also change the hydrology of soils, sometimes making them water-repellent, increasing runoff and erosion when rains return. The aftermath can shift vegetation composition: grasslands can replace forests, changing future fire regimes. These slower feedbacks are less cinematic than a pyroCb, but they set the stage for the next extreme.

Clouds remain central because they are the atmosphere's fastest lever on surface energy and moisture. Smoke influences clouds not only by seeding droplets but also by changing how sunlight heats the land, which changes convection. A smoke-dimmed surface may not warm enough to break a morning inversion, allowing low clouds or fog to persist longer. That can reduce daytime heating further. Conversely, absorbing smoke aloft can create a warm layer that caps the boundary layer, trapping heat and humidity near the surface overnight, sometimes worsening heat stress. People living under smoke frequently report odd temperature sensations: cooler afternoons but oppressive nights, or a day that feels dim yet stifling. Those impressions are not imagination; they are a human-scale rendering of a vertical redistribution of energy.

The climate-modeling community has become increasingly attentive to these couplings because cloud feedbacks already dominate much of the uncertainty in projected warming, and fire is one more source of aerosol and one more disruptor of cloud regimes. Some of the latest generations of global climate models have shown higher effective climate sensitivity than earlier ones, and stronger cloud feedbacks are part of the explanation. If low-cloud changes amplify warming, and warming amplifies fire weather, and fires modify aerosols and clouds, then fires become entangled with the same cloud uncertainties that shape the broad projections. The danger is not that models miss every detail of fire—cloud interaction, but that missing details could tilt the sign or magnitude of regional responses, especially in places where people live and where ecosystems are already stressed.

It is tempting, amid these loops, to search for a simple managerial solution: could smoke be "useful" as a cooling agent, a kind of accidental geoengineering? The idea surfaces periodically, often implicitly, when people note that aerosols cool and that smoke plumes can be large. But wildfire smoke is an uncontrolled, toxic, and temporally erratic aerosol source. Its composition varies with fuel type and combustion conditions; its injection height varies with meteorology and fire intensity; its health impacts are immediate and severe. Even if one were willing to entertain deliberate aerosol interventions in the atmosphere, the governance literature has emphasized transparency, coordination, and inclusive engagement precisely because the risks and side effects are unevenly distributed. Wildfires offer a grim preview of what "unplanned intervention" looks like: communities suffering hazardous air, ecosystems damaged, and the atmosphere perturbed in ways that are hard to predict and harder to reverse.

A better way to understand smoke's climatic role is as a stress test of the planetary thermostat. In an ideal thermostat, a warming perturbation triggers responses that cool, damp, and restore balance. Fire adds both damping and amplification pathways. Shading and cloud brightening can cool locally, briefly easing heat. Rainfall suppression, drying, and lightning-driven ignitions can accelerate the very conditions that produce more fire. Stratospheric smoke can persist and reshape radiation over large scales, nudging circulation and cloud patterns far from the flames. The sign of the net effect is not fixed; it depends on the details of aerosols, clouds, and weather, and on the timescale of interest.

On the ground, the most consequential feedback is often the simplest: heat and dryness make fuels receptive, and wind delivers oxygen and spreads embers. Yet above the flames, the atmosphere is not merely carrying smoke away. It is reacting, adjusting, and sometimes returning the favour. A pyroCb anvil that shades the fire line for an hour can also send gusty outflows that drive the front into new terrain. A smoke layer that dims the Sun can also warm the air aloft and choke off the small convective clouds that would otherwise have wrung moisture from the boundary layer. A cloud deck seeded by smoke can brighten and cool, yet the missing drizzle can leave forests a little thirstier next week.

In that sense, wildfire smoke is both mirror and misdirection: a brightening of clouds here, a drying of the land there; a temporary cooling at noon, a lingering warmth at night; a plume that looks like weather but behaves like climate. The sky above a large fire is a laboratory no one asked for, running experiments at full scale, with results written in ash on windshields and in thin brown layers drifting, quietly, at the edge of space.

Chapter 7

Walking the Tightrope

Earth's climate stability is not a static fortress but a precarious balancing act constantly under siege by amplifying forces. This chapter investigates the fragile boundaries of this equilibrium, exploring how positive feedbacks can trigger irreversible changes and why our planet narrowly escaped the frozen fate of Mars or the inferno of Venus.

7.1 The Physics of No Return

Some changes in climate are like turning a dimmer switch: nudge the dial, the light brightens; nudge it back, the room returns to where it was. Other changes behave more like a light with a sticky button: press once and it clicks on; pressing again does not undo the first click in the same smooth way. The unsettling feature is not merely speed. It is asymmetry. The route into a new state can be easy, while the route back can be steep, narrow, or blocked altogether.

Physicists have a name for this asymmetry: *hysteresis*. The word comes from the Greek for "lagging behind," and it first became famous in nineteenth-century studies of magnetism, where iron retains magnetization after the external field is reduced. Climate hysteresis is the same idea written large: the state of the system depends not only on the conditions *now* but on the path taken to get here. In a hysteretic system, there can be more than one stable climate under the same external forcing. Which one you occupy depends on history.

A simple way to picture this is a ball rolling on a landscape. Valleys are stable states: nudge the ball and it rolls back. A ridge between valleys is a threshold: push the ball high enough and it tumbles into a different basin. On the far side, you can lower the push and the ball still stays put, because it is now cradled by a new valley. Returning it to the original valley requires not merely reversing the original push, but supplying enough energy to climb back over the ridge. That is the physics behind the phrase "point of no return" when it is used carefully. It does not mean that nature flips a moral switch and forbids recovery. It means recovery demands a different, harder physical journey.

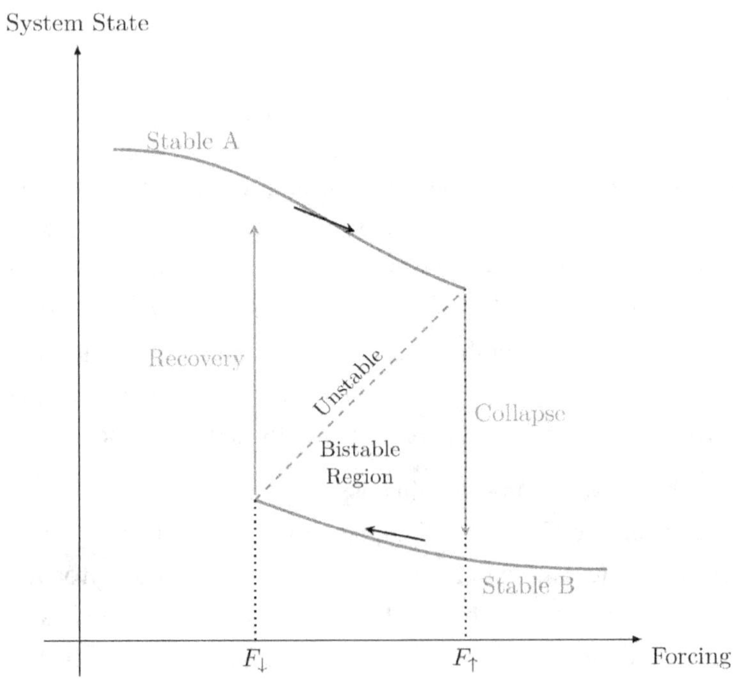

The curve in the diagram is the abstract skeleton of hysteresis. Imagine the horizontal axis as some external "push" on climate: greenhouse gases, sunlight, freshwater input to the ocean, anything that changes the energy or buoyancy balances. The vertical axis is a state variable: global ice cover, ocean circulation strength, forest extent. For a range of forcing, two stable branches exist: the system can sit in a high-state or a low-state. As forcing increases gradually, the system can track the upper branch until it reaches a critical point, marked $F_{\uparrow}$, where that stable solution vanishes. At that moment there is nothing to "hold" the ball. It falls to the lower branch. If you then reverse the forcing, the system does not jump back at $F_{\uparrow}$, because the lower branch remains stable there. It only returns when forcing is reduced far enough to hit a different threshold, $F_{\downarrow}$. The gap between these thresholds is the hysteresis loop: a memory of the path.

This is why "reversibility" in climate needs to be handled with care. A temperature number alone is not a magic key; what matters is whether the system has left the basin of attraction that made the old climate self-repairing. In human terms, the problem is not simply that it might get warmer than we would like. The problem is that warming can rearrange the internal plumbing of the planet in ways that make the old arrangements difficult to restore even if temperatures later fall.

The underlying mathematics often begins with a particular kind of instability called a *saddle-node bifurcation*. That phrase can sound like a trapdoor into jargon, but the idea is plain. Picture a shallow valley that slowly flattens as the external push increases. A ball in that valley returns to the bottom after a shove. But as the valley flattens, the restoring force weakens: nudges take longer to damp out. Eventually, the valley disappears entirely, leaving only a slope. At that point the ball slides away. A system poised near such a threshold often exhibits *critical slowing down*: it becomes sluggish at recovering from disturbances. In data, this can show up as increased "stickiness" in year-to-year changes, like a swing that takes longer to settle after a push.

The temptation is to treat critical slowing down like an alarm bell: measure rising variance and autocorrelation in a time series, and you can anticipate the fall. Sometimes this works in controlled experiments and in relatively clean ecological records. In the climate system it is trickier. The world is noisy, observations are short, and the forcing itself is changing rapidly. A patient can show worrying symptoms that mean different things depending on context; climate indicators behave similarly. The absence of an early-warning signal does not guarantee safety, and the presence of one does not prove a cliff is ahead. That uncertainty is not comforting, but it is honest.

There is another twist that makes "no return" less like a line on a map and more like a moving target: the *rate* at which forcing changes can matter as much as the final value. If a system has an internal timescale—the time it takes to adjust—and you change the environment faster than that, the system can be tipped without ever crossing the static threshold you would calculate in a textbook. It is like jerking a steering wheel: you can spin out even on a road that is, in principle, navigable. This kind of rate-driven tipping matters in climate because industrial society has become a forcing machine with an unusually fast hand on the dial.

To see why timescale matters, it helps to return to the ball. Suppose the valley is slowly being deformed while the ball rolls. Even if the valley has not vanished, the ball may be left behind on a part of the landscape that is no longer a true minimum because it cannot keep up with the moving bottom. Real climate components—ice sheets, forests, deep ocean circulation—have inertia. Ice sheets have to flow; trees have to grow; the ocean has to mix. The existence of inertia does not make them safe. It means their response can lag, and lag can hide commitment. By the time the change is obvious, the ball may already be on the slope.

A historical example makes this more concrete, and it comes from a moment when people realized the planet could behave in jumpy ways even without human forcing. In the early twentieth century, a young Serbian mathematician and engineer named Milutin Milankovitch worked out how subtle changes in Earth's orbit alter the distribution of sunlight over seasons and latitudes. His calculations were elegant, but for decades they were mainly a theoretical curiosity. Then, in the 1970s, geochemists studying deep-sea sediment cores—notably the work associated with James Hays, John Imbrie, and Nicholas Shackleton (1976)—showed that ice-age rhythms in oxygen isotopes lined up strikingly with Milankovitch's orbital cycles. The story was not simply "sunlight changes, ice changes." The sunlight variations were too small to explain the size of ice-age swings on their own. The missing amplification was feedback: ice reflects sunlight, and greenhouse gases respond to temperature and ocean circulation. In other words, the climate system had internal levers that could magnify a gentle orbital nudge into a global transformation.

Those ice-age records also hinted at hysteresis. The growth of continental ice sheets and their collapse do not mirror each other perfectly. Ice can advance gradually as summers cool and snowfall persists, but collapse can be abrupt when melting and ice dynamics reach a point where the structure destabilizes. The exact details differ among ice sheets and intervals, but the general lesson is robust: a climate component can sit in a stable configuration for a long time, then change quickly once a threshold is crossed. The return journey is not simply the movie played backward.

For a more immediate and anxiously discussed case, consider the Atlantic Meridional Overturning Circulation, often shortened to AMOC. This is not a single current but a pattern: warm surface waters flow northward, release heat to the atmosphere, become denser, sink, and return southward at depth. It is one reason northwestern Europe is milder than its latitude might suggest, and it is tightly linked to rainfall patterns on both sides of the Atlantic. The crucial physical ingredient is density: seawater becomes denser when it is colder and saltier. Add fresh water from melting ice or increased rainfall, and you make surface waters lighter. Light water resists sinking. If sinking weakens, the northward flow of warm water can weaken as well, which can further cool the region and alter ice and freshwater inputs. A loop is born.

In simplified models of the ocean, the AMOC can have two stable states under the same forcing: a strong overturning mode and a weak mode. That is the hysteresis picture drawn in the diagram, translated into seawater. Push the system with enough freshwater in the right places, and the strong circulation can shut down or reorganize. Remove the freshwater forcing, and the circulation does not necessarily snap back at once; the density structure has changed, the pathways of mixing have changed, and the ocean's memory is long. Some of the most vivid warnings about abrupt Atlantic change come not from modern instruments but from paleoclimate: during the last glacial period, Greenland ice cores record abrupt warmings and coolings on the order of decades, known as Dansgaard–Oeschger events. Their causes remain debated in detail, but changes in Atlantic circulation are widely considered a major player. The uncomfortable takeaway is not that tomorrow the Atlantic will flip like a switch; it is that a circulation system capable of multiple modes has a built-in mechanism for surprise.

Modern assessments are careful: they typically judge an abrupt AMOC collapse before 2100 as unlikely under many scenarios, while still emphasizing deep uncertainty and potentially severe impacts. That mixture—low probability, high consequence—is precisely the kind of risk that hysteresis makes hard to handle psychologically. People are good at reacting to trends; they are less good at living with thresholds that may or may not be near.

Ecosystems can carry hysteresis in a different guise: not through ocean density, but through biology's own feedback loops. The Amazon rainforest, for instance, helps make its own climate. Trees transpire water; that moisture feeds clouds and rainfall; rainfall supports trees. Disturb the forest by drought, heat, deforestation, or fire, and you can weaken the moisture recycling that sustains it. Grasslands and degraded forests reflect and absorb energy differently, burn differently, and create different atmospheric conditions. The system can become stuck in a drier state even if the original trigger is reduced, because the very machinery that produced humidity has been dismantled. This is not mysticism about "nature balancing itself." It is physics and ecology interacting: energy, water, and carbon are being rerouted.

The key phrase here is *basin of attraction.* A stable climate state is not a single point but a region: a set of conditions from which the system tends to return to its familiar configuration after perturbations. Hysteresis is the geometry of those basins and the ridges between them. Once the system crosses a ridge, the same small restoring forces that once pulled it home now pull it toward a different home. In everyday life, there are analogous situations: trust can be stable until it is broken, and then rebuilding requires more than undoing the original offense. But in climate the "more than" is not moral effort; it is energy and time.

Energy matters because many climate thresholds are tied to feedbacks that alter the planet's energy balance. If melting sea ice exposes dark ocean, more sunlight is absorbed, warming accelerates, and more ice melts. The reverse process requires not only cooling but creating the conditions under which ice can persist through summer, rebuild thickness, and restore its reflective shield. Time matters because some parts of the climate system integrate change. The deep ocean stores heat for centuries. Ice sheets respond over centuries to millennia. Carbon cycles through fast and slow pathways. A small mismatch between incoming and outgoing energy—a few watts per square meter, the power of a dim light bulb spread over a tabletop—accumulates into vast amounts of stored heat when given enough years.

The phrase "point of no return" is often used casually to mean "after this, things get bad." Hysteresis gives it a sharper, and sometimes more troubling, meaning. There can be a *commitment* point and a *realization* point. Commitment is when the ball has crossed the ridge in a practical sense: even if you stop pushing, it will still roll down into the new valley because momentum and landscape guarantee it. Realization is when the ball reaches the bottom and the consequences become obvious to an observer. The gap between them can be long. In ice sheets, for example, the geometry can be committed to retreat while the visible mass loss unfolds over centuries. That gap is one reason climate risks can feel abstract until they are suddenly not.

It is also why early-warning signs, even when real, can be socially difficult to act on. Imagine being told that the system is slowing down in its recovery from perturbations, that variance is rising, that the ridge may be near. None of this tells you exactly where the ridge is, and none of it guarantees that the next push will be the one that tips the ball. Yet waiting for certainty is a strategy that hysteresis punishes. Once the threshold is crossed, the cheapest actions are no longer available.

There is a deeper, quieter lesson in the physics. Stabilizing feedbacks—the planetary thermostat idea in its broad sense—work by providing restoring forces. They are not invincible. A restoring force can be weakened, delayed, or overwhelmed by self-reinforcing loops. In a hysteretic system, the stabilizers that hold one state in place may not exist, or may operate differently, in the alternative state. A thermostat can keep a room comfortable only if the furnace and the air conditioner remain functional and if the walls stay intact. Burn down the house, and the thermostat can still click, faithfully, uselessly.

The point is not despair. The point is discrimination: to learn which parts of Earth's behavior are dimmer switches and which have sticky buttons. Some features of climate are relatively reversible on human timescales once forcing is reduced; others are not. Some thresholds are sharp; others are fuzzy. Some changes are gradual in the global average but abrupt regionally, where people live. Hysteresis tells us that the most dangerous stories are not always the ones with the steepest trend lines. They are the ones where the line looks gentle until the landscape underneath changes shape.

A planet that can flip between stable states is not a planet out of control. It is a planet with multiple modes of control, each with its own rules. Human activity is now a forcing strong enough to test the ridges between those modes. The tightrope is not stretched over a single abyss. It is stretched over a terrain of hidden saddles and competing valleys, where the safest step is the one that keeps the ball well away from the crest.

7.2 The White Earth Paradox

Imagine stepping outside at noon and feeling almost no warmth from the Sun. Not because the Sun has dimmed, but because the planet has turned into a mirror. Snow and ice do not merely signal cold; they help make it. Fresh snow can reflect most incoming sunlight back to space, so a patch of white ground is not just a consequence of winter but a small machine that deepens winter's grip. Give that machine enough territory, and it can begin to run itself.

The Neoproterozoic Era, roughly 1,000 to 541 million years ago, appears to have hosted episodes when this machine nearly took over the entire Earth. Geological clues scattered across continents point toward glaciations so extreme that ice may have reached the tropics, perhaps even the equator. The phrase *Snowball Earth* is vivid for a reason: it captures the possibility that the oceans froze over on a global scale, leaving a world of white continents, white seas, and a thin band of survival wherever liquid water could persist.

At first glance, this seems to contradict the very idea of a planetary thermostat. If the climate system contains stabilizing feedbacks, why would it allow itself to fall into such a deep freeze? The paradox is sharpened by another puzzle: once a planet becomes very white, how does it ever thaw? A frozen Earth should reflect sunlight so efficiently that warming seems almost impossible. Yet the rocks suggest not only that the planet escaped, but that it may have rebounded violently into a hot aftermath.

The key to the paradox lies in time. A thermostat can fail in the short term even if it succeeds in the long term, especially when the stabilizing mechanisms operate slowly and the destabilizing ones act quickly. Ice—albedo feedback is fast. The carbon cycle, which can eventually rescue the climate, is usually slow. When those timescales are separated widely enough, the planet can be knocked into a stable but extreme state—and then hauled out again by a mechanism so gradual that it feels almost like fate.

The mechanics begin with a simple piece of physics: Earth's energy budget. The planet's temperature settles where incoming solar energy absorbed by Earth balances outgoing heat radiated to space. Albedo—the fraction of sunlight reflected—controls the first half of that balance. Increase albedo, and less solar energy is absorbed, so the planet cools. Cooling expands snow and ice, which increases albedo further. That is the classic positive feedback. In the 1960s, Mikhail Budyko showed with simple energy-balance ideas that this feedback can produce more than one stable climate state: a warm state with limited ice and a cold state with much more extensive ice. Between them lies a threshold where small changes in forcing can produce a jump.

The Neoproterozoic seems to have wandered close enough to such a threshold for the feedback to run away. The world's continents were arranged differently then; the Sun was slightly fainter than today; and the chemistry of the atmosphere and oceans was not the same as in the recent geological past. None of those details alone forces a snowball. What matters is that the combination may have made the warm state less strongly anchored. Once ice reached low enough latitudes, the feedback could take over: brighter Earth, less absorbed sunlight, more cooling, more ice.

The evidence for low-latitude glaciation is not one spectacular smoking gun but a pattern that looks strange if read through the lens of modern climate. Geologists find glacial deposits—rocks dropped by ice, scratched bedrock, debris flows associated with glaciers—in places that, when reconstructed on ancient maps, appear to have been near the equator. If those reconstructions are right, glaciers existed where today one expects rainforests. There are also layers known as *banded iron formations* reappearing around some of these intervals, hinting at oceans that became unusually stratified and oxygen-poor beneath an ice cover. Then, abruptly above certain glacial sequences, there appear distinctive carbonate layers called *cap carbonates*: thick blankets of limestone or dolostone laid down in conditions that seem to have shifted suddenly away from the icy regime.

A historical anecdote captures how unsettling this pattern was when it was pulled together into a coherent story. In the late 1990s and early 2000s, field geologists working in Namibia, Australia, and elsewhere began to connect these disparate clues into a single hypothesis. Paul Hoffman and Daniel Schrag (2002) argued that the simplest way to explain tropical glacial deposits and the curious cap carbonates was to accept the extreme: the planet had frozen over almost entirely. It was a bold claim because it forced a re-evaluation of what the climate system is capable of, and it demanded a physical mechanism for escape.

Escape is where the paradox becomes sharpest. A mostly ice-covered Earth is highly reflective. Under a bright white shell, the ocean is insulated from direct sunlight, and the atmosphere above is cold and dry. One might imagine the planet locked in permanent winter, a stable end state. Yet the rocks say otherwise: life survived, and the planet thawed.

The rescue mechanism traditionally invoked is volcanic CO_2. Volcanoes do not stop erupting because the planet is frozen; the interior heat engine continues. Over millions of years, volcanoes and metamorphic processes add CO_2 to the atmosphere. Under ordinary conditions, CO_2 is drawn down by chemical weathering of silicate rocks on land. Rainwater mixed with CO_2 becomes weakly acidic and reacts with minerals, ultimately storing carbon in carbonate rocks and carrying it to the ocean. This weathering thermostat is a negative feedback: warmer temperatures and more rainfall speed weathering, pulling CO_2 down and cooling the planet; colder temperatures slow weathering, allowing CO_2 to accumulate and warm the planet.

A snowball changes the rules. If continents are buried under ice and the hydrological cycle is stunted, the usual pathways for silicate weathering weaken dramatically. The thermostat that normally prevents runaway warming is effectively switched off, not because it has failed as a principle, but because its physical gears are jammed by ice. CO_2 can then build up in the atmosphere, slowly but relentlessly, because the source continues while the sink is muted.

Eventually, the greenhouse effect strengthens enough to overcome the high albedo. At some threshold CO_2 concentration, the planet warms, melting begins, and once darker surfaces are exposed, the ice—albedo feedback runs in reverse: less ice lowers albedo, increasing absorbed sunlight and accelerating warming. The termination can be abrupt compared with the long frozen interval. A world that took a long time to enter the snowball might exit rapidly, not because volcanism suddenly accelerates, but because the system crosses a threshold where feedbacks flip sign.

This neat story is compelling, and it may be broadly correct, but it is not the end of the science. The details matter because they determine how hard it is to deglaciate a frozen planet and what the aftermath looks like. Raymond Pierrehumbert's modeling work in the mid-2000s showed that a "hard snowball"—a planet with global sea ice and a largely shut-down ocean-atmosphere exchange—can be stubbornly stable. In some simulations, deglaciation required very high CO_2 levels, and the outcome was sensitive to cloud behavior and the way heat is transferred near the surface. That sensitivity is not an embarrassment; it is a warning that the threshold depends on real atmospheric physics, not just on hand-waving energy balance.

It also matters because the aftermath of a snowball is expected to be dramatic. If CO_2 truly built up for millions of years with little weathering sink, the atmosphere at deglaciation could have contained orders of magnitude more CO_2 than preindustrial levels. The instant the ice retreats, rainfall resumes, and warm conditions accelerate weathering. The old thermostat comes back online and begins to pull CO_2 down. In the classic picture, this drawdown is slow, governed by the pace at which continents weather and the ocean buries carbonates: perhaps hundreds of thousands of years of greenhouse heat after a rapid thaw.

Cap carbonates fit naturally into that classic picture: they look like the geological signature of a world suddenly flooded with carbonate-saturated waters and intense chemical weathering. But cap carbonates themselves have become part of the paradox, because their textures, chemistry, and distribution hint at processes more complex than a simple, slow weathering drawdown.

Recent work has sharpened this debate by focusing on the ocean. The ocean is not a passive bucket waiting for rivers to deliver chemistry; it has its own carbon-cycle dynamics, circulation patterns, and chemical equilibria. After a global glaciation, the ocean may have been strongly stratified, with different layers holding different amounts of dissolved carbon and alkalinity (the capacity of seawater to neutralize acids and support carbonate formation). When the ice retreats, meltwater freshens the surface, potentially maintaining stratification for a time, while air—sea gas exchange begins again. Under such conditions, CO_2 can shift between ocean and atmosphere on timescales much shorter than continental weathering.

Modeling efforts in 2024, including work by Ramme and Ilyina, suggested that atmospheric CO_2 after deglaciation might not simply decline slowly and monotonically. Depending on initial ocean chemistry and disequilibria, the ocean could draw down atmospheric CO_2 rapidly on timescales shorter than 10,000 years, or in other cases could outgas CO_2 and intensify warming temporarily. That is a profound change in emphasis. It does not remove the importance of silicate weathering as a long-term stabilizer, but it adds a fast, ocean-driven layer to the story that can reshape the climatic whiplash at the end of a snowball.

Cap carbonates are central here because they are the mineral record of carbonate chemistry in action. A particularly detailed 2024 proposal by Thomas and Catling framed cap carbonate formation as a multi-stage process involving both syn-glacial and post-glacial chemistry: alkalinity could build up in the deep ocean through low-temperature reactions even during the glaciation; after deglaciation, intense continental weathering and meltwater could create a shallow, carbonate-saturated layer that precipitates carbonates rapidly; mixing with deeper alkalinity could prolong deposition. The details are still contested—cap carbonates differ from place to place, and no single model explains every feature—but the direction of travel is clear. The end of a snowball is not just a slow return of rain and rivers; it is a reawakening of a chemically and physically reorganized ocean-atmosphere system.

There is an emotional element to this story that the rocks cannot quite convey, but that becomes inescapable when one tries to imagine living on such a planet. The surface would have been hostile: cold, dry air; limited open water; fierce winds sweeping over ice. Yet microbial life persisted, and perhaps pockets of more complex life survived in refuges—near geothermal heat, in cracks and thin-ice regions, or in open-water "polynyas" kept from freezing by winds and currents. The Snowball hypothesis gained traction partly because it forced biologists and geochemists to ask a new kind of question: not simply how climate changes, but how life endures when climate locks into an extreme state for millions of years.

One of the most striking human elements comes from the way the idea itself was built: by people walking across deserts and mountains, reading ancient climates in scattered outcrops. In Namibia's Skeleton Coast region, for example, geologists have traced glacial deposits and cap carbonates across long distances, mapping the transitions like detectives reconstructing a crime scene from footprints. The work is slow, hot, and often uncomfortable—an irony, considering it is used to infer global cold. A Snowball Earth is not an image born in a computer. It is an image assembled from hammer blows, compass bearings, and the stubborn habit of asking why one layer of rock sits so strangely atop another.

The white Earth paradox, then, is not simply “how could it freeze?” and “how could it thaw?” It is also: how could the same feedbacks that amplify change also permit long-term recovery? The answer is that feedbacks do not come labeled good or bad. Ice—albedo feedback is powerful precisely because it is a physical property of water and sunlight. It does not care whether it is helping a stable climate maintain polar caps or helping a glaciation spread toward the equator. The carbon-cycle thermostat is stabilizing precisely because it responds to temperature and rainfall—but those are the very things a snowball suppresses, temporarily disabling the stabilizer.

A useful way to hold the pieces together is to think in terms of competing clocks. The ice—albedo feedback can run on the timescale of seasons to centuries. Ocean circulation can reorganize on decades to centuries, though the full deep ocean adjustment takes longer. Volcanic CO_2 input accumulates on hundreds of thousands to millions of years. Silicate weathering, in the long-term sense that balances volcanic input, is similarly slow. When the fast loop pushes hard enough, it can carry the system into a different basin of attraction. When the slow loop finally builds pressure in the other direction, it can push the system out again—sometimes with a snap, because the fast loop is still there, waiting to accelerate whichever direction the threshold dictates.

This is why Snowball Earth is so valuable as a case study for a planet with a thermostat. It demonstrates that thermostats can have operating ranges. Beyond those ranges, the same physics that normally damps change can be turned off, and the same amplifiers that usually act locally can become global. It also demonstrates that recovery can be both inevitable and violent. A frozen world can be rescued by a slow accumulation of greenhouse gases, yet the act of rescue can overshoot, producing a climate shock in the opposite direction.

A planet that can swing between these extremes is not fragile in the sense of being easily destroyed. It is fragile in a more subtle sense: it can reorganize itself into states that are stable, persistent, and profoundly unfamiliar. The rocks do not show a gentle return to normality; they show abrupt boundaries, strange chemical signatures, and the imprint of a world that once became its own mirror and then shattered that mirror in a rush of heat and rain. The white Earth paradox is that a feedback strong enough to imprison the planet in ice is also strong enough to break the prison quickly once the lock finally gives way.

7.3 The Venusian Cautionary Tale

Venus hangs in the evening sky like a promise: bright, serene, almost gentle. For most of human history it was the planet of beauty. The shock is not that Venus is different from Earth—all planets are—but that it is different in the particular way that matters most for habitability. Venus is close in size and bulk composition, probably formed from similar building blocks, and sits only one orbit inward from Earth. Yet its surface is hot enough to melt lead, its air is thick enough to crush, and its clouds are made of sulfuric acid. If Earth has a thermostat, Venus is what it looks like when that thermostat is broken or bypassed.

A thermostat is not a single device. It is an ensemble of feedback loops that, taken together, tend to oppose runaway change: water moving heat around, clouds reflecting sunlight, rocks reacting with CO_2, oceans absorbing carbon, and ice brightening or darkening the planet. The two previous stories—a hysteresis landscape with ridges and valleys, and a snowball world rescued by slow carbon accumulation—share a theme: Earth can be knocked far from its familiar state, yet it has routes back because certain stabilizing processes remain available. Venus, by contrast, is a lesson in what happens when the stabilizers depend on ingredients that can be lost.

The most important of those ingredients is ordinary, unglamorous water.

Water does two jobs at once. As a greenhouse gas, water vapor traps infrared heat, warming the surface. As a condensing substance, water also rains out, forming oceans that can dissolve gases, erode rocks, and lubricate a long carbon cycle. These two roles pull against each other. On Earth, water vapor is a powerful amplifier of warming: warmer air holds more moisture, so warming tends to increase water vapor, which warms further. Yet water does not accumulate indefinitely in the atmosphere because it condenses and falls as rain or snow. That condensing behavior is a kind of governor. It keeps atmospheric water vapor tied to temperature rather than to an ever-growing reservoir.

The Venusian caution begins with a subtle point in radiation physics: there is a limit to how much heat a wet atmosphere can radiate to space. The planet cools by emitting infrared radiation, but in a very moist atmosphere the infrared-emitting level rises to high altitudes where temperatures are constrained by atmospheric structure. As more water vapor is added, the atmosphere becomes more opaque in the infrared, and the outgoing heat does not increase as quickly as one might expect. In certain conditions it approaches a ceiling. If the absorbed sunlight exceeds that ceiling, the planet cannot shed enough heat to balance what it receives. The surface warms, evaporating more water, which further thickens the greenhouse blanket. This is the core idea of a *runaway greenhouse*: a self-reinforcing transition where the condensed reservoir (an ocean) can be driven into the atmosphere.

That phrase is sometimes used loosely as a synonym for “very warm.”

Physically, it means something sharper: radiative equilibrium becomes impossible with liquid water on the surface. The climate cannot find a stable temperature while keeping oceans. If pushed hard enough, it must move into a qualitatively different state.

There is a related but distinct pathway called the *moist greenhouse*. Here, the planet may still maintain surface liquid water for some time, but the upper atmosphere becomes wet. On Earth today, the stratosphere is relatively dry because water vapor condenses out before air reaches those heights. A moist greenhouse breaks that dryness: more water reaches the stratosphere, where sunlight can split water molecules, and the hydrogen can escape to space. Water is then lost not as steam that later rains out, but as an actual depletion of the planet's inventory. One-dimensional climate modeling under idealized assumptions has long suggested that a full runaway greenhouse for an Earth-like planet might require something like 1.4 times the modern Earth solar flux, while moist greenhouse water loss could become efficient at lower forcing in some scenarios. The exact numbers shift with clouds, atmospheric composition, and circulation, but the qualitative warning is robust: there are climates in which water stops being a temporary atmospheric guest and starts becoming a one-way export.

Venus receives about twice as much sunlight as Earth, simply because it orbits closer to the Sun. That does not guarantee a runaway greenhouse by itself; the outcome depends on albedo, clouds, rotation, and initial conditions. But it places Venus in a different regime of risk. If early Venus had oceans—a possibility still debated but not ruled out—it would have been easier for warming to drive more water vapor into the atmosphere, pushing toward either a moist greenhouse that bleeds water away or a full runaway that evaporates the oceans.

The harshest part of the Venus story is that once the oceans are gone, the carbon thermostat that stabilizes Earth has nowhere to stand. On

Earth, CO_2 is regulated over long times by a dance between volcanic outgassing and rock weathering. Weathering is powered by liquid water: rain, rivers, and the chemical reactivity of wet minerals. The products are carried to the ocean, where carbonates form and are buried, locking carbon into rock. Plate tectonics and metamorphism then return some carbon to the atmosphere through volcanism, completing the cycle. It is not perfect, and it is not quick, but it is a feedback loop with a sign: more warmth tends to mean more weathering, which tends to draw down CO_2 and cool the planet.

Remove liquid water and you remove the loop's lever arm. A dry planet can still have chemical reactions and can still bury carbon in some forms, but the efficient global conveyor that moves carbon from air to rock slows to a crawl. CO_2 then becomes less a dial and more a ratchet. If volcanism continues, atmospheric CO_2 can build up and remain high, and the atmosphere becomes a long-lived heat trap.

Venus's present atmosphere is overwhelmingly CO_2, at a pressure roughly 90 times that of Earth's atmosphere. That mass of gas is itself a clue: Earth and Venus likely started with comparable supplies of carbon, but Earth stores most of its carbon in rocks (limestones, sediments, and altered crust), while Venus stores much more in the atmosphere. The simple contrast is not that Venus has a greenhouse effect and Earth does not; Earth's greenhouse effect is essential to habitability. The contrast is that Earth has multiple reservoirs and pathways that move carbon between them. Venus looks like a planet where carbon has been stranded in the air.

A concrete human story helps make this less abstract. In December 1978, NASA's Pioneer Venus probes dropped into the Venusian atmosphere, measuring its structure and chemistry as they descended. A few years earlier, in 1970, the Soviet Venera 7 had transmitted the first data from the surface of another planet, surviving only minutes in the crushing heat. Those missions were feats of engineering and courage by proxy: the willingness to send instruments into a place designed to destroy them. They gave a new kind of intimacy with Venus. It was no longer a bright point in the sky or a speculative canvas. It had temperature profiles, pressure readings, chemical ratios, and clouds with a specific composition. Planetary cautionary tales became empirical.

Among the measurements that loom largest in Venus's water story is the ratio of deuterium to hydrogen, often written D/H. Deuterium is "heavy hydrogen," with an extra neutron. When water molecules are broken apart and hydrogen escapes to space, the lighter ordinary hydrogen escapes more readily than deuterium. Over time, this preferential loss leaves behind a higher proportion of deuterium in the remaining water. Venus's atmosphere is strongly enriched in deuterium relative to Earth. That enrichment is consistent with substantial hydrogen escape in the past, and therefore substantial water loss.

But here the caution needs a second layer. A high D/H ratio is not a simple ruler that tells you how many Earth oceans Venus once had. It provides constraints and hints, not a unique biography. The inference depends on assumptions about the planet's initial water inventory, the history of volcanic outgassing, the efficiency of escape at different times, and whether water was replenished or sequestered. Multiple evolutionary histories can fit the same present-day ratio. Scientific caution here is not fence-sitting; it is an acknowledgment that planetary history is written in partial sentences. Venus's D/H enrichment strongly suggests that water was lost and that escape mattered, but it does not uniquely prove that Venus began as a twin of Earth with a global ocean of the same scale.

Even with those uncertainties, Venus still serves as a stringent boundary condition for thinking about Earth. The reason is that Venus embodies a set of coupled feedbacks that, once triggered, become hard to reverse.

The coupling begins with water vapor. As a greenhouse gas, it amplifies warming. Warming then allows more water vapor, strengthening the greenhouse effect. But the deeper coupling is between water and carbon. If warming and radiation physics push enough water into the upper atmosphere, hydrogen escape can remove water irreversibly over geological times. As oceans shrink, the capacity to dissolve CO_2 declines, and the chemical weathering sink weakens. As CO_2 rises, the greenhouse effect strengthens, pushing temperatures higher and making it even harder for liquid water to persist. A planet can then slide into a state where the very processes that would cool it require the ingredient that has been lost.

This is "no return" in its most literal physical sense: not just hysteresis in temperature, but hysteresis in planetary inventory. A snowball Earth can thaw because the carbon cycle continues to supply greenhouse gas even when weathering is shut down, and because water is still present in some form, ready to restart the chemical machinery once temperatures rise. Venus suggests a different kind of trap: lose enough water, and the chemical machinery cannot restart even if forcing changes later. The thermostat is not merely overwhelmed; it is dismantled.

It is tempting, and common, to frame Venus as a moral: a warning label for Earth's future. That framing can mislead if it implies a simple equivalence. Earth is farther from the Sun, receives less stellar energy, and has a different atmospheric history. The timescales of human-caused warming are also vastly shorter than the geological processes that likely shaped Venus. No serious climate scientist claims that burning fossil fuels will turn Earth into present-day Venus in any foreseeable timeframe.

Yet Venus remains relevant because it clarifies the limits of feedback thinking. If the stabilizing loops rely on physical components that can be depleted—liquid water, exposed silicate rocks, a functioning ocean circulation—then stability is conditional. Under certain forcings, the climate system does not merely shift along a continuum; it can cross into regimes where entirely different constraints apply. That is what makes Venus unsettling. It is not that Venus is Earth's fate. It is that Venus shows the existence of climates where the usual escape routes are gone.

The most sobering implication is psychological as much as physical. People often hear "feedback" and imagine a self-correcting system, like a thermostat that prevents extremes by design. But feedback can also mean self-destruction when the loop's sign is positive and the limiting processes are removed. A thermostat is a device with a goal only because humans built it with one. Planets have no goals. They have couplings. Earth's couplings happen to allow a long-lived habitable window, but that window is not guaranteed by benevolence; it is maintained by a fortunate arrangement of physics and chemistry.

Venus is sometimes called Earth's "evil twin," a phrase that flatters our sense of drama. The real lesson is less theatrical and more intimate. Earth's livable climate is not maintained by one heroic mechanism, but by many small ones that share a few fragile dependencies: liquid water that can cycle between sky and sea, rocks that can react, and time enough for slow chemistry to matter. On Venus, those dependencies appear to have been severed. The result is a planet where the greenhouse effect is not a protective blanket but a permanent weight, pressing down with the accumulated consequences of an early loss.

In the evening, Venus is still beautiful. It still shines with borrowed sunlight. Its brightness is an albedo trick: the planet is wrapped in reflective clouds. The irony is sharp. Venus is bright not because it is cold, but because its atmosphere is thick and clouded, reflecting sunlight while trapping heat below. The same basic physics—reflection and radiation—that helps make Earth temperate can, under different circumstances, make a world both dazzling and uninhabitable.

Chapter 8

Hand on the Dial

For millions of years, Earth's temperature was regulated by slow geologic grinds and subtle orbital shifts, but now a single species has seized the controls. We examine the physics of this sudden takeover, revealing how the sheer speed of human emissions has overwhelmed the planet's natural dampening systems. The narrative concludes with the daunting reality of thermal inertia and the technological gamble of attempting to artificially engineer the atmosphere.

8.1 The Overwritten Signal

There is a particular kind of quiet in the numbers that come out of ice. When scientists drill down through the Greenland or Antarctic ice sheets, each meter brings them deeper into compressed snowfall laid down long before cities, agriculture, even before our species. Tiny bubbles trapped in that ice are samples of ancient air, sealed and stored like messages. When the bubbles are opened in the lab and analyzed, the past speaks in a calm, repetitive cadence: carbon dioxide rises a little, falls a little, over and over, paced by slow shifts in Earth's orbit and the advance and retreat of ice.

If the modern atmosphere could be bottled in the same way, it would not read like another verse in that long song. It would look like someone abruptly seized the pen and scribbled across the page.

Rate is the heart of the disturbance. Climate history is full of change, some of it dramatic. But much of Earth's built-in machinery depends on time: processes that act like brakes on carbon dioxide and temperature are powerful in the long run, yet sluggish on the scale of a human life. For most of the past million years, the key climate signals have been gentle enough that these brakes could keep up. Today the forcing is arriving so quickly that the old signals are still there, still operating, but they are increasingly drowned out—not because they have stopped, but because we have turned up the volume on something else.

A useful way to grasp this is to imagine two clocks ticking in the same room. One is a grandfather clock: steady, slow, and loud in its own register. The other is a digital alarm that suddenly starts beeping. The grandfather clock keeps perfect time; its tick has not changed. The room simply becomes dominated by the alarm.

Milankovitch cycles are the grandfather clock. They are small, predictable changes in Earth's orbit and tilt that redistribute sunlight across seasons and latitudes. On their own, they do not add energy to the climate system; they reshuffle it. Yet over tens of thousands of years they can tip the balance between a world that grows ice sheets and a world that melts them. The trick is that the reshuffling nudges feedbacks—especially ice reflectivity and greenhouse gases—that amplify the initial push. It is a slow conversation between sunlight, ice, oceans, and air.

Industrial-era emissions are the alarm. They change the *composition* of the atmosphere quickly, making it harder for Earth to radiate heat to space. That is not a reshuffling of sunlight; it is a change in the planet's heat-loss plumbing. The difference is crucial. A slight orbital nudge can matter if it persists long enough for the carbon cycle and ice sheets to respond. But a rapid injection of carbon can matter even if the orbits never budge, because it pushes directly on the greenhouse effect at a pace the slower parts of the system cannot immediately counter.

One of the most vivid comparisons comes from a natural event often described as a stress test for Earth's thermostat: the Paleocene—Eocene Thermal Maximum, usually shortened to PETM. Around 56 million years ago, global temperatures rose sharply in geological terms. The deep ocean warmed, ecosystems shifted, and the carbon cycle was violently disturbed. The PETM is not a perfect analogue for today—the continents were arranged differently, there were no polar ice sheets to melt, and the cause of the carbon release is still debated. Yet it provides something rare: an episode where nature itself dumped a great mass of carbon into the air—ocean system and then spent a very long time cleaning it up.

The detail that matters most for our purpose is not merely that carbon dioxide rose, but how quickly. Even if the PETM carbon release unfolded rapidly by the standards of sediment layers, the best evidence suggests it still spread over thousands of years, not decades. That gap—thousands versus decades—is the difference between a push the Earth system has time to partially absorb and a punch that lands before the body can brace.

A chart helps because the human mind is not built for logarithms of time. We feel the difference between a minute and an hour; we struggle with the difference between a century and ten millennia. The comparison below is deliberately stylized: it does not pretend to reconstruct the PETM in exact parts per million, because the uncertainties are large and the proxies indirect. Instead it aims to show the central point about *rate*: the modern rise is steep on a timescale that makes even the PETM look gradual.

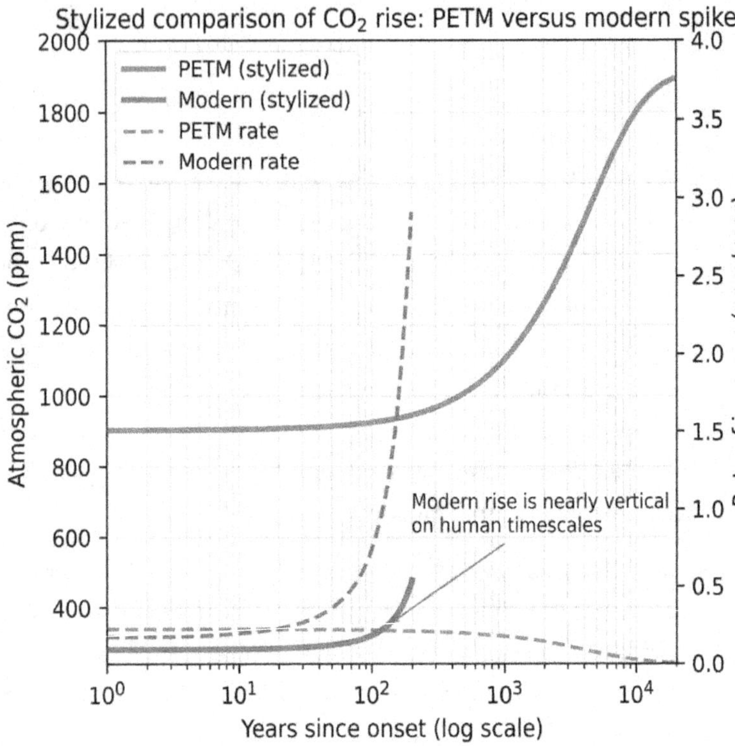

The dashed curves in that figure are the quiet revelation: the rate of increase during the modern episode is measured in whole parts per million per year, while the rate during the PETM is better thought of as hundredths of a part per million per year when averaged over its long ramp. The precise ratio depends on what one assumes about the PETM's timing and magnitude, but the order-of-magnitude contrast is robust: modern carbon release is extraordinarily fast.

This is where a historical anecdote becomes more than a charming story. In 1958, a geochemist named Charles David Keeling began measuring atmospheric carbon dioxide at the Mauna Loa Observatory in Hawaii. The site was chosen because it sits high enough above the local vegetation and pollution to sample air that represents a broad swath of the Pacific. Keeling's measurements did something that ice cores could not: they captured change in real time. The record quickly revealed two patterns. One was the seasonal breathing of the Northern Hemisphere's land plants—CO_2 falling in spring and summer as leaves draw it down, then rising in autumn and winter as decay returns carbon to the air. The other was a steady upward climb that did not reverse when seasons changed.

The seasonal cycle was an education in humility: the planet's biosphere moves carbon around so energetically that it leaves a fingerprint detectable in the global atmosphere. But the long-term climb was the alarm. It showed that human emissions were not merely perturbing a local air mass; they were shifting the entire baseline of the atmosphere itself. Decade by decade, the curve rose. From the perspective of the ice-core record, it was as if a slow tide had been replaced by a rapidly filling bathtub.

Here the word *signal* is not a metaphor but a technical description. Climate scientists talk about signals and noise the way engineers do. Weather is noisy: temperatures rise and fall from day to day, year to year, with storms, droughts, ocean cycles, volcanic eruptions. A climate signal is a persistent pattern that can be distinguished from this variability. Milankovitch forcing is a signal, but it is faint. It is measurable, calculable, and in the long run profoundly important, yet on any given decade it is barely a whisper compared with the noise of weather.

The modern greenhouse signal has become loud enough to be measured directly in the planet's energy flows. Think of Earth as a budget with two major columns. Income is solar energy absorbed by the planet. Expenses are heat radiated back to space, mostly as infrared. When income and expenses balance, global temperature is roughly stable. When income exceeds expenses, the planet accumulates energy. That accumulation is not an abstraction. It is the extra heat stored in the ocean, the energy melting ice, warming land, and changing the atmosphere. A positive imbalance sustained year after year is like a slow but relentless deposit into a savings account you did not intend to open.

The phrase *top-of-atmosphere energy imbalance* names this difference between absorbed sunlight and outgoing heat. Satellites can measure it by observing reflected sunlight and emitted infrared radiation. The measurements are not trivial; they require careful calibration and cross-checking with what the ocean is doing. But the headline is simple: the imbalance is positive. Earth is taking in more energy than it is losing.

This is one of the key ways the human forcing overwrites the natural rhythm. Orbital cycles redistribute sunlight, but they do not create a persistent global energy surplus in the same direct way. Greenhouse gases do. They act like adjusting the thickness of the blanket, altering the efficiency with which the planet sheds heat. When that blanket thickens quickly, the planet is pushed out of balance faster than many of its slow stabilizers can respond.

Natural feedbacks still operate, and some of them respond quickly. Water vapor, for example, amplifies warming because warmer air holds more moisture, and water vapor itself is a greenhouse gas. Changes in clouds can either amplify or damp warming depending on type, altitude, and location; cloud feedbacks are among the most studied and debated details in climate science precisely because they matter so much and are so complex. Ice and snow respond by retreating, reducing reflectivity and allowing more sunlight to be absorbed—a clear amplifying feedback, especially in the Arctic.

But the most relevant stabilizers for carbon dioxide are slower, and they depend on chemistry and geology. The ocean absorbs CO_2 from the atmosphere, and on timescales of decades to centuries it can take up a large fraction of what we emit. Yet that uptake is limited by circulation: much of the ocean is isolated from the surface for long periods, so the deep sea cannot immediately participate in the exchange. On still longer timescales, the Earth system has an even deeper brake: chemical weathering. Rainwater reacts with rocks; the products of that reaction eventually carry carbon to the ocean, where it can be locked away in carbonate minerals. This is the thermostat's long-term negative feedback. It is powerful enough to stabilize Earth's climate over millions of years, but it is not fast enough to prevent a sharp spike on the timescale of industrial society.

The consequence is that the modern signal does not merely ride atop the natural cycles; it dominates them. One way to see this is to remember what the natural range looked like just before the industrial era. Over many thousands of years, CO_2 drifted within a relatively bounded envelope, rising during warm interglacials and falling during ice ages. The changes were large enough to matter, but slow enough for ecosystems to migrate and for the ocean and biosphere to adjust. In contrast, modern CO_2 has climbed far beyond that recent envelope in a handful of generations. The rapidity is part of the danger. Biological and social systems are adapted not only to a climate, but to the pace at which climate changes.

That pace has a human face. Consider the story of the steam engine and coal. In the early nineteenth century, Britain's coal-powered industry began to scale in a way that transformed economies and landscapes. The power was intoxicating because it substituted ancient sunlight, stored in fossil carbon, for muscle, wind, and flowing water. Railways expanded, iron production surged, cities thickened with soot. The fuel that made this possible was not just energy-rich; it was carbon-rich. When coal is burned, the oxygen that combines with carbon ends up as CO_2, and the molecule is stable enough to persist and accumulate.

The key is accumulation. If the atmosphere were a river that instantly flushed away everything poured into it, emissions would matter far less. Instead, the atmosphere is more like a lake with an outlet: it mixes quickly, but it drains slowly. Plants and the ocean draw down some of the added CO_2, yet the drawdown is incomplete on human timescales. That is why the Keeling Curve keeps rising even as forests grow in some regions and oceans absorb vast amounts of carbon. The sinks are working; the sources are simply overwhelming them.

Once this is understood, the idea of an overwritten signal becomes less mysterious and more physical. The natural signals have characteristic frequencies: tens of thousands of years for orbital variations, centuries to millennia for deep-ocean circulation adjustments, many millennia for weathering feedbacks to bite. A forcing that arrives over a century is like striking a bell with a hammer rather than stroking it with a bow. The tone changes. The system responds in a different register, dominated by fast feedbacks and constrained by slow ones.

The rewriting is also evident in the atmosphere's bookkeeping of oxygen and carbon isotopes, which act like forensic markers. Fossil fuels are ancient plant matter, depleted in one of carbon's heavier isotopes. When they are burned, they leave a distinctive isotopic signature in atmospheric CO_2. Meanwhile, combustion consumes oxygen. High-precision measurements have detected both patterns: the isotopic shift and the small decline in atmospheric oxygen consistent with widespread burning. These are not arguments by analogy; they are fingerprints.

Yet the most consequential part of the overwritten signal is not that it proves causation. It is that it changes what the future must contend with. When Earth's climate is nudged slowly, the slow feedbacks participate in the nudge. When the nudge becomes a shove, the slow feedbacks lag behind, and the system can be pushed into states that would have been unlikely under the old rhythm.

There is an everyday analogy that does not trivialize the physics. Imagine steering a large ship. If you turn the wheel gently, the ship responds slowly but smoothly, and you can correct as you go. If you spin the wheel rapidly, the ship still responds slowly—it cannot do otherwise—but now you have created a mismatch between input and response. By the time the ship begins to turn, you may already have overcorrected. In the climate system, the mismatch between the speed of our forcing and the speed of the system's largest components is a central source of risk.

It is tempting, at this point, to search for reassurance in the idea that Earth has been warmer before, that CO_2 has been higher before, that life has survived worse. All of that is true in a narrow sense, and yet it misses the essential point. The question is not whether Earth can be warm. The question is whether a civilization built around coastlines, stable growing seasons, reliable water supplies, and inherited infrastructure can remain resilient when the background conditions are changing faster than many natural and human systems are accustomed to adapting.

The overwritten signal is, finally, a change in what counts as "normal." A generation ago, a warming trend could still be argued about as if it were merely a statistical curiosity, something that might vanish into the noise. Now the planet's energy budget has tilted, and the atmosphere's composition has shifted enough to be read as a direct, persistent forcing. The old natural metronomes continue to tick, but a new beat has been imposed upon them—one that does not fade at the end of a season, and does not wait for the slow turning of the celestial gears.

8.2 The Thermal Debt

A pot of water on a stove teaches a lesson that feels almost unfair the first time you burn your tongue. You turn off the flame and expect the danger to vanish. It does not. The water keeps steaming; the metal stays hot; the temperature lingers above what is safe. The heat you added is not instantly undone by the simple act of stopping.

The climate system has the same stubbornness, but on a planetary scale. Carbon dioxide can stop rising, or even begin to fall, and the planet can still keep warming for a while. Not forever, not without limits, but long enough to matter to coastlines, crops, and lives. The lag is not a political talking point or a quirk of computer models. It comes from something as ordinary as heat capacity: the ability of a substance to soak up energy without changing temperature very quickly. Water has a famously large heat capacity. The oceans cover about 70% of Earth's surface. Put those together and you get a kind of thermal flywheel: once it is spinning, it resists sudden changes.

It is tempting to picture global warming as a thermostat in a house: you turn the dial up, the air gets warmer; you turn the dial down, it cools again. But a house is mostly air, thin and quick to respond, and its walls are not three kilometers deep. A closer comparison would be a massive building with a heated concrete foundation. Turn down the furnace and the air cools first, but the foundation continues to release heat into the rooms for hours or days. In Earth's case the "foundation" is the ocean, and the timescale is not hours or days but decades and longer.

One way to put the idea in plain language is to call it *thermal debt*. Borrowing money feels painless at the checkout counter; the cost shows up later, with interest. Adding greenhouse gases is similar: the atmosphere changes quickly, but the full response of the planet is delayed. The debt is the warming and sea-level rise already "in the pipeline" because the ocean has been absorbing heat faster than it can give it back. Even if the borrowing stops, the interest payments continue for a while.

The physics begins at the top of the atmosphere, where sunlight arrives and heat leaves. When greenhouse gases increase, the planet becomes slightly less efficient at radiating infrared energy to space. That creates an *energy imbalance*: more energy coming in than going out. The imbalance is small compared to the total flow of solar energy, which is why it is easy to underestimate. Yet spread over the surface of the Earth and sustained year after year, even a fraction of a watt per square metre is enormous. It is like a tiny leak in a dam that, given time, fills a vast reservoir.

Where does this extra energy go? Not mainly into the air. The atmosphere is thin. If Earth's energy gains were stored primarily in the air, the temperature would lurch upward much faster. Instead, most of the excess is absorbed by the ocean. The Intergovernmental Panel on Climate Change assessed in its 2021 physical-science report that roughly 91% of the additional energy accumulating in the Earth system in recent decades has gone into the ocean. Land, melting ice, and the atmosphere itself make up the remaining small fractions. The ocean is doing the heavy lifting, and that is both a blessing and a trap. It blunts immediate surface warming, but it also stores heat that cannot be wished away.

It helps to imagine the ocean as layered. The top layer—the mixed layer—is stirred by winds and waves and exchanges heat with the atmosphere relatively quickly. Below that lies the thermocline, a zone where temperature changes rapidly with depth, which acts as a barrier to mixing. Beneath that, the deep ocean is cold, vast, and sluggish. Heat can reach it, but it must be carried by slow processes: the sinking of dense water in certain regions, the broad conveyor-belt circulation, and mixing driven by tides and currents rubbing against rough seafloor.

Because of this layering, the ocean responds like a system with multiple gears. The surface can warm in a season; the upper few hundred metres can shift over years to decades; the deep ocean can keep absorbing heat for centuries. When the surface of the ocean warms, it does not merely track the air temperature; it becomes a heat reservoir that interacts with the atmosphere, shaping storms, humidity, and rainfall. When the deeper ocean warms, it commits the planet to long-term change, especially in sea level.

The relationship between emissions and temperature is often discussed as if it were a direct line: emit less, warm less. Over the long run, that is broadly true. But in the near term, the relationship has a lag and a twist. Consider an idealized thought experiment: suppose humanity could freeze greenhouse gas concentrations at their current level. Would warming stop immediately? No. The planet would still be out of balance: it would still be receiving more energy than it radiates to space. As long as that imbalance persists, the ocean will keep taking up heat. Surface temperature would keep rising until the planet warmed enough that outgoing infrared radiation once again matched incoming solar energy. That is equilibration, and it takes time.

Now consider an even more dramatic experiment: suppose emissions stop abruptly. Concentrations would not stop instantly, because carbon dioxide is not like smoke that clears overnight; it is a long-lived gas with a slow drawdown. David Archer (2009) emphasized how misleading it is to speak of a single "lifetime" for CO_2: some fraction is removed on timescales of decades to centuries, but a substantial tail remains for many centuries and longer because the ultimate removal involves slow ocean chemistry and even slower geological processes. A multi-model analysis by Joos and colleagues (2013) captured the same idea in impulse-response curves: a pulse of carbon leads to a rapid rise in temperature, followed by only a slight decline over centuries to a millennium, even as the ocean continues to take up heat.

This is the uncomfortable meaning of thermal debt: there is a delay between changing what we *do* (emissions) and changing what the planet *feels* (temperature and its many consequences). The delay is not uniform. Some impacts respond quickly. Heatwaves and heavy rainfall can shift within years as the atmosphere warms and holds more moisture. Other impacts are slow but relentless. Sea level is the clearest example because it integrates the ocean's heat content over time.

To see how the lag works in everyday terms, consider the historical lesson of Mount Pinatubo. In June 1991, the volcano erupted in the Philippines and lofted a veil of sulfur aerosols into the stratosphere. Those tiny particles reflected sunlight back to space, temporarily cooling the planet. Global average surface temperature dipped for a year or two. This event is often used to test climate models because it is a natural experiment in altering Earth's energy budget.

What is less often appreciated is the ocean's role in how quickly the cooling showed up and how the climate system recovered. The atmosphere responded quickly to the reduced sunlight, but the ocean's massive heat capacity moderated the dip and helped govern the return toward the prior trend as the aerosols settled out. Pinatubo was a brief brake applied to a system already being pushed by greenhouse gases. The fact that the surface responded within a couple of years shows that the climate system can be sensitive. The fact that the long-term trend resumed underscores that a short interruption does not erase a persistent energy imbalance.

Another concrete example, closer to human experience, comes from the ocean itself: marine heatwaves. In the northeast Pacific from

2013 to 2016, a blob of unusually warm surface water appeared and persisted, disrupting ecosystems and fisheries. Although this event had specific weather and ocean-circulation causes, it was made more damaging by the fact that baseline ocean temperatures have risen. A warmer ocean increases the odds that a natural fluctuation will cross a threshold that ecosystems cannot tolerate. The broader warming of the ocean provides the background; the weather supplies the spark.

Thermal debt also changes how we should interpret year-to-year variability at the surface. People sometimes point to a few cool years, or a plateau in the rate of surface warming, as evidence that the problem has paused. But the planet's energy imbalance does not pause simply because the surface temperature has a slow year. The ocean can take up more heat during some periods and release more during others, influenced by El Niño and La Niña cycles, volcanic eruptions, and fluctuations in clouds. Surface temperature is the part we live in, but it is not the entire ledger. The deeper ledger is the ocean's heat content, and it tends to rise steadily when the top-of-atmosphere imbalance remains positive.

The metaphor of a bank account becomes useful again. Imagine your paycheck increases, but your visible spending does not change much for a while because you are quietly paying down an old loan, or building savings. Looking only at what is in your wallet on a given day might mislead you about your financial trajectory. The ocean is that hidden account. When it gains heat, it is storing the excess energy imposed by greenhouse gases even if the atmosphere's warming temporarily slows.

The debt is not only thermal; it is also mechanical. Warm water expands. This is called *thermal expansion*, and it is a major contributor to sea-level rise. Even if ice sheets and glaciers stopped melting tomorrow, sea level would continue to rise for some time as the ocean slowly warms at depth. Thermal expansion is a quiet kind of change: it does not require a dramatic collapse or a visible break. It is simply physics, operating in a volume so immense that millimetres accumulate into centimetres and then into tens of centimetres.

Then there is the other part of sea-level rise: ice. The ocean's heat does not remain politely in the ocean. It reaches the margins of ice sheets and glaciers, thinning floating ice shelves and warming fjords where glaciers meet the sea. This is where the debt becomes visceral. Communities do not experience "ocean heat content" directly. They experience higher tides reaching farther onto streets, saltwater intruding into wells, storm surges riding on a higher baseline. A positive energy imbalance is a technical phrase; a flooded subway entrance is a human one.

There is a subtlety here that deserves care. The lag does not mean that stopping emissions is pointless or that nothing improves for centuries. Some things improve quickly. Air pollution from burning fossil fuels includes particles that harm health and also reflect sunlight; reducing fossil fuel use can rapidly improve air quality and save lives within months to years. The rate at which temperatures rise can slow relatively quickly if emissions fall sharply, because the growth of atmospheric CO_2 slows and methane can decline faster. Over longer spans, stabilizing temperature requires reaching net-zero CO_2 emissions, and then potentially net-negative if one aims to lower concentrations.

But the presence of thermal debt means that there is no instant off-switch for the impacts already set in motion. It also means that there is a difference between *stopping the rise of the forcing* and *reversing the forcing*. A stabilization of greenhouse gases halts the worsening of the energy imbalance, but it does not instantly restore balance. Returning the planet to balance requires either reducing greenhouse gases enough for outgoing heat to catch up, or changing incoming sunlight, or some combination. The ocean sits in the middle of this story as both buffer and liability.

The carbon-cycle tail makes this even more consequential. Christine Kaufhold and Matteo Willeit (2025) argued, using carbon-cycle modelling, that while much anthropogenic CO_2 can be removed over centuries to millennia after emissions cease, a small residual fraction can persist for very long times, longer than 100,000 years. The exact numbers depend on the size of the emission pulse and the details of ocean chemistry, land carbon storage, and weathering. The message is not that catastrophe is fixed for geological time, but that the climate system has memory. Our actions can write themselves into the atmosphere and ocean long after their immediate benefits were enjoyed.

The emotional weight of thermal debt comes from its mismatch with human instincts about cause and effect. People are used to feedback that is immediate: touch a hot pan and you recoil; lift your foot off the accelerator and the car slows. Climate feedback is delayed and distributed. The costs show up in places far from the smokestack, and in decades far from the decision that created them. The lag can make it psychologically easy to postpone action, because the worst consequences seem to belong to someone else's future.

A historical analogy from outside climate can sharpen the point. In the nineteenth century, some cities built water and sewer systems only after repeated cholera outbreaks forced the issue. The engineering was not mysterious; the delay came from politics, cost, and the difficulty of acting on an invisible threat. Once the infrastructure was built, the benefits were profound and lasting, but so were the consequences of having waited. Thermal debt has a similar structure: the system keeps a record. The question is not whether the record exists; it is how large it becomes before the inputs change.

There is also a caution embedded in the idea of debt: paying it down is harder than avoiding it. Preventing a tonne of CO_2 from entering the atmosphere is generally easier than removing that tonne later, because removal requires work—capturing dilute molecules and storing them safely for long periods. Preventing heat from entering the ocean is easier than cooling the deep ocean, because cooling requires a sustained negative energy imbalance, which is not something the planet naturally supplies on demand.

Still, the physics offers a kind of clarity. When emissions fall, the rate at which the debt grows can slow. When net-zero is reached, the *principal* of the debt can stop increasing. If concentrations are then reduced, the terms of the loan change: the energy imbalance can shrink, the warming can stabilize, and in time some surface temperatures can ease. The ocean, however, will not forget quickly. It will continue to adjust, to mix, to expand and circulate, carrying the imprint of decades of imbalance.

It is worth lingering on what "adjust" means in physical terms. Heat does not merely sit passively in seawater. It changes density, and density drives currents. Warm water is generally less dense than cold water, which can strengthen stratification—the layering that inhibits mixing. Stronger stratification can reduce the ocean's ability to draw down carbon by limiting the contact between the surface and the deep. It can also change nutrient supply to the surface, with consequences for marine ecosystems. In this way, thermal debt is not just a delayed temperature response; it is a restructuring of the ocean's internal dynamics.

A person standing at the shoreline, watching waves break, could be forgiven for thinking the ocean is nothing but restless motion. Yet the largest part of the ocean's motion is slow, deep, and persistent, like breathing in a sleeping giant. Thermal debt is heat stored in that slow motion, and it will be paid out over time: as continued sea-level rise from expansion, as altered currents and weather patterns, as a baseline shift that makes extremes more likely. The ocean has been quietly absorbing the consequences of our altered energy budget, and it will continue to speak that absorbed history back to the atmosphere for a long time.

8.3 Hacking the Thermostat

Once the idea of a planetary energy budget becomes familiar—sunlight in, heat out, the difference stored mostly in the ocean—a reckless thought arrives with a certain logic. If the problem is an imbalance, why not fix the balance directly?

In ordinary life, we do this all the time. When a room is too warm, we do not negotiate with the laws of thermodynamics; we open a window, switch on a fan, or adjust the thermostat. The climate system is not a room, and there is no dial on the wall, but the underlying temptation is understandable: if greenhouse gases are turning down Earth's ability to shed heat, perhaps we can either remove those gases or compensate by reflecting a little more sunlight.

Those are the two great families of intentional climate intervention. They are often grouped under the uneasy umbrella word *geoengineering*, which can mean everything from planting forests to putting particles in the stratosphere. A more careful distinction, used by the Royal Society (2009) and later by the U.S. National Academies (2021), separates interventions by what part of the energy budget they target.

Carbon Dioxide Removal (CDR) aims to reduce atmospheric CO_2 by taking it out of the air (or preventing it from entering) and storing it for long periods. In energy-budget terms, CDR tries to restore the planet's ability to radiate heat to space by thinning the greenhouse blanket. It addresses the root cause of much warming and, crucially, it also addresses ocean acidification, because acidification is driven by CO_2 dissolving into seawater, not by temperature alone.

Solar Radiation Management (SRM) aims to reduce the amount of sunlight absorbed by the Earth, typically by increasing reflection (albedo) or altering clouds. In energy-budget terms, SRM tries to reduce the incoming side of the ledger. It can, in principle, cool the surface relatively quickly, but it does not remove CO_2, so it does not solve acidification, and it introduces a different category of physical and political risk.

A schematic makes the difference concrete, in the simple language of arrows.

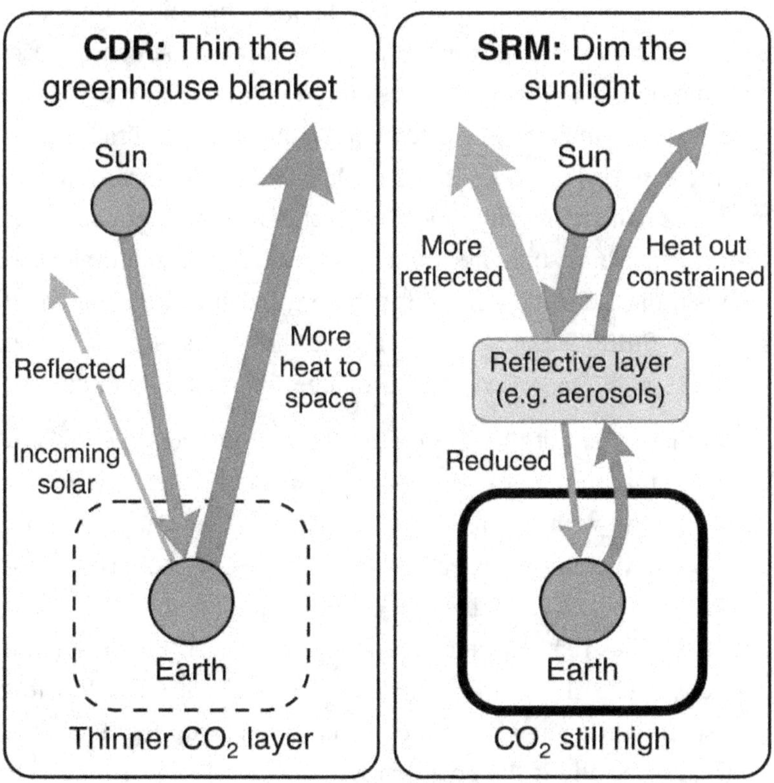

The simplicity of the picture is part of its danger. It can make interventions look like interchangeable knobs: either widen the heat-out arrow (CDR) or shrink the sun-in arrow (SRM). In reality, these options differ not only in mechanism but also in timescale, side effects, and governance.

CDR is slow because it must fight the scale of the atmosphere. Air is vast and CO_2 is dilute. Pulling molecules out of that mixture takes energy and infrastructure. Nature does some of the work already: plants, soils, and the ocean absorb roughly half of human CO_2 emissions each year, though not in a way that can keep concentrations from rising when emissions remain large. CDR proposes to increase the removal rate beyond what nature is currently doing.

Some forms of CDR are familiar and can be beneficial for reasons beyond climate. Reforesting a degraded landscape can store carbon in wood and soils while restoring habitat and improving local water cycles. Improving soil carbon through agricultural practices can increase resilience to drought. Other forms are more industrial: machines that chemically bind CO_2 from air (direct air capture), or systems that burn biomass for energy and then capture and store the resulting CO_2 underground (bioenergy with carbon capture and storage). Still others involve enhancing the ocean's uptake through chemical additions or accelerated weathering of certain rocks.

The difficulty is not that these are physically impossible. The difficulty is that they must be done at enormous scale, reliably, over long periods, without creating new harms. Underground storage must remain secure for decades and centuries. Land-based carbon storage can be reversed by fire, pests, or policy changes. Ocean-based schemes risk altering ecosystems that we still understand imperfectly. The slow carbon-cycle tail means that CDR has to be persistent: removing CO_2 is not a one-off cleanup but a sustained project, especially if the goal is not merely to slow warming but to reverse some of it.

SRM looks temptingly different. It does not need to remove CO_2 molecule by molecule. It aims to change Earth's reflectivity by a small fraction, enough to reduce absorbed sunlight and cool the surface. Because sunlight is so powerful, a small percentage change can have a large temperature effect. This is why SRM is often described as fast: if you could brighten the planet, cooling could follow within months to a few years, roughly the timescale on which the atmosphere adjusts.

The most discussed SRM approach is *stratospheric aerosol injection* (SAI), which would deliberately place reflective particles high in the stratosphere. The basic physical idea is inspired by volcanoes. When Mount Pinatubo erupted in 1991, sulfur dioxide gas reached the stratosphere and formed sulfate aerosols that reflected sunlight. The planet cooled measurably for a short period. That natural event serves as both proof-of-concept and warning label: proof that aerosols can cool, warning that aerosols affect more than temperature.

Other SRM concepts include *marine cloud brightening* (MCB), which would spray sea-salt particles into low marine clouds to make them more reflective, and *cirrus cloud thinning* (CCT), which aims to reduce the warming effect of high, thin cirrus clouds that tend to trap outgoing infrared radiation. The National Academies' 2021 report treated these as research targets rather than deployment plans, emphasizing the need for governance and careful study precisely because the uncertainties and risks are large.

One of the most sobering aspects of SRM is how directly it intervenes in patterns of sunlight and rainfall. Climate is not merely a global average temperature; it is the distribution of heat, moisture, and circulation. If SRM reduces sunlight, it can cool the planet, but it can also change the hydrological cycle. Less sunlight at the surface generally means less evaporation. Less evaporation can mean changes in precipitation patterns, potentially affecting monsoons and regional water availability. The same global mean temperature can be achieved by different mixes of greenhouse forcing and solar reduction, and those mixes do not necessarily produce the same regional climates.

Marine cloud brightening illustrates how tricky this becomes. Clouds are not simple mirrors; they are living microphysical systems, built from droplets and aerosols, sensitive to updrafts, humidity, and background pollution. Daniel Wood (2021) emphasized that MCB efficacy depends strongly on meteorology and the existing aerosol environment. A cloud over a pristine ocean may respond differently than a cloud already influenced by ship exhaust or continental pollution. What works in one region or season could fail in another. The very thing that makes MCB seem attractive—its potentially regional, adjustable nature—also makes it difficult to predict and govern.

There is also the problem of *mismatch*. SRM can reduce temperature, but it does not remove CO_2. That means it does not fix ocean acidification, which threatens coral reefs and shell-forming organisms by altering seawater chemistry. It also means that the atmosphere remains loaded with greenhouse gas, still pushing down on outgoing infrared radiation. In the schematic, the greenhouse "box" stays thick. SRM is best understood as masking some symptoms while leaving the cause in place.

Masking can be lifesaving in medicine; it can also be dangerous if it discourages the treatment of the underlying disease. This is where SRM's political and psychological risks come into view. If a technology can cool the planet quickly, it may create incentives to delay the hard work of cutting emissions and scaling CDR. This concern is often called *moral hazard*, though the phrase can sound accusatory. The deeper point is structural: a fast, visible lever can crowd out slower, less glamorous investments, even when the slow investments are the ones that actually solve the problem.

Then there is the sharpest edge of the SRM risk profile: *termination shock*. If SRM is deployed to offset a large amount of greenhouse warming and then stops abruptly—because of political conflict, economic collapse, technical failure, or a decision that the side effects are intolerable—the planet would experience a rapid warming as the masked greenhouse forcing reasserts itself. Andy Parker and Peter Irvine (2018) explored how this risk is often discussed in dramatic physical terms while the probability pathways, governance robustness, and failure modes receive less careful attention. The physics is straightforward: if you have been holding down a spring and suddenly let go, it snaps back. The social difficulty is ensuring that a system intended to run continuously for decades can actually be maintained through wars, regime changes, and shifting public consent.

A small, real-world episode hints at what termination can feel like, even without any deliberate SRM program. On 1 January 2020, new international rules sharply reduced sulfur content in marine fuel, cutting sulfur dioxide emissions from shipping. Sulfur aerosols from ships had been providing a modest cooling effect by brightening clouds and scattering sunlight, especially over busy shipping lanes. When emissions fell, that inadvertent cooling diminished. Researchers have examined the resulting radiative changes as a kind of accidental analogue: not a global SRM shutdown, but a sudden reduction in aerosol masking over parts of the ocean. The episode is valuable not because it proves a single, clean temperature signal—the climate system is noisy—but because it demonstrates how quickly aerosol forcing can change, and how tightly that forcing is intertwined with air quality, regulation, and politics.

Aerosols also raise a set of ethical trade-offs that are difficult to avoid. The same particles that cool can harm. Sulfate aerosols contribute to respiratory illness and acid rain when they are in the lower atmosphere. Stratospheric aerosols avoid some of those direct health impacts near the surface, but they bring other concerns, including potential impacts on ozone chemistry and the whiteness of the sky. Deciding that one set of risks is acceptable in exchange for reduced heat extremes is not merely a technical decision. It is a decision about whose risks count, who gets to decide, and who bears unintended consequences.

This is why governance is not an afterthought but central to the question. The National Academies (2021) argued for a coordinated research program and research governance, distinguishing research from deployment. Even research can be politically charged, because knowledge is not neutral in its consequences. A credible SRM research program would need transparency, international engagement, and mechanisms for public accountability. It would also need humility about what models can and cannot resolve. The history of environmental interventions is full of confident engineering that later revealed hidden couplings: pesticides that reshaped ecosystems, river dams that altered sediment and fisheries, invasive species introduced with good intentions.

The fundamental asymmetry remains. CDR is slow, expensive, and likely to be limited in how quickly it can scale, but it works in the same direction as the planet's long-term stabilizers: it reduces atmospheric CO_2 and therefore reduces radiative forcing and acidification. SRM is comparatively fast in its temperature effect, but it is a substitute signal imposed on top of a still-perturbed carbon cycle. It asks humanity not simply to stop breaking the thermostat, but to take over its operation.

There is a sobering clarity in the energy-budget arrows. Widening the heat-out arrow by removing CO_2 is difficult because it requires years of construction, policy, and careful accounting, and it must be sustained long enough to matter. Shrinking the sun-in arrow may be easier in a narrow technical sense, but it commits the world to a continual act of maintenance, with uncertain regional outcomes and the constant danger that stopping quickly would produce a surge of warming.

A civilization can manage many kinds of long-term projects, but it struggles when the project requires uninterrupted global cooperation and when failure carries a rapid, shared penalty. The atmosphere does not care whether the interruption is accidental or deliberate. If the masking ends, the underlying forcing remains, and the planet continues to collect its due.

www.ingramcontent.com/pod-product-compliance
Lightning Source LLC
LaVergne TN
LVHW022211200926
848481LV00006B/1931
* 9 7 9 8 8 9 6 6 5 2 8 1 6 *